Assessment and Renovation of Concrete Structures

Concrete Design and Construction Series

SERIES EDITORS

PROFESSOR F. K. KONG
University of Newcastle upon Tyne

EMERITUS PROFESSOR R. H. EVANS CBE
University of Leeds

OTHER TITLES IN THE SERIES

Concrete Radiation Shielding: Nuclear Physics, Concrete Properties and Construction *by M. F. Kaplan*

Quality in Precast Concrete: Design — Production — Supervision *by John G. Richardson*

Reinforced and Prestressed Masonry *edited by Arnold W. Hendry*

Concrete Structures: Materials, Maintenance and Repair *by Denison Campbell-Allen and Harold Roper*

Concrete Bridges: Design and Construction *by A. C. Liebenberg*

Assessment and Renovation of Concrete Structures

Ted Kay

(with Travers Morgan Consultants)

Longman
Scientific &
Technical

Copublished in the United States with
John Wiley & Sons, Inc., New York

Longman Scientific & Technical,
Longman Group UK Limited,
Longman House, Burnt Mill, Harlow,
Essex CM20 2JE, England
and Associated Companies throughout the world.

Copublished in the United States with
John Wiley & Sons, Inc., 605 Third Avenue, New York, NY 10158

First published in 1992

British Library Cataloguing in Publication Data
A catalogue entry for this book is available from the British Library.

ISBN 0582-05779-5

Library of Congress Cataloging-in-Publication Data
A catalogue entry for this book is available from the Library of Congress.

ISBN 0470-21864-9

Set by 4Z in 10/12 pt Times
Printed and Bound in Great Britain
at the Bath Press, Avon

Contents

Foreword   ix

Acknowledgements   xi

1   Introduction   1
    Constituents   2
    Manufacture of Portland cement   2
    Portland cement composition   6
    Other cementing materials   10
    Aggregates   13
    Water   18
    Admixtures   19
    References   24

2   Deterioration Processes   25
    Construction and design defects   25
    Early-age defects   26
    Longer term problems   31
    References   47

3   Planning an Investigation   49
    Safety   49
    Overall plan   51
    Initial general phase   51
    Detailed phase   53
    Commissioning testing work   65
    References   66

4   Testing Techniques   68
    Strength testing   68
    Covermeter (pachometer)   86
    Half-cell potential   89
    Resistivity   91

Radar   92
Dust sampling   93
Carbonation depth testing   95
Radiography   95
Strain   96
Permeability   98
Laboratory test methods   101
Alkali reactivity   107
References   108

5   Interpretation of Results   111
Strength tests   113
Structural assessment   121
High alumina cement concrete construction   126
Load tests   127
Chloride contents   128
Carbonation   131
Carbonation and chlorides   133
Sulphate contents   136
Half-cell potential results   136
Alkali — silica reactivity   138
References   143

6   Repair and Renovation Techniques   145
Preliminary considerations   145
Sprayed concrete   147
Patch repair   152
Bulky repairs   157
Crack filling and resin injection   160
Bonding of external reinforcement   164
References   165

7   Protection   166
Introduction   166
Surface treatments   167
Cathodic protection   176
Chloride extraction and replenishment of alkalis   187
References   189

8   Contract Documents   191
Introduction   191
Specification for the work   193
Concrete Society patch repair specification   196
Dutch repair recommendations   197

Sprayed concrete   204
Department of Transport specifications   204
Cathodic protection   207
References   209

Index   211

Foreword

One of the privileges of being President of The Institution of Structural Engineers, as I was in 1987−88, is the opportunity created by the requirement to travel widely both in the United Kingdom and overseas. The purpose is in part to visit members in their own branches and countries but also to represent the Institution at international gatherings of engineers. In both instances the President is able to visit work in the course of construction, some of which is straightforward and some of which is of high technical merit and content. What struck me repeatedly was the very high volume of work which related to existing buildings, some of obvious historical or cultural interest, but most emanating from the vast development programmes, often in hostile environments, of the last 30 years or so.

Whilst in some instances it was apparent that the fabric was suffering from marked deterioration, in others it was simply that the first period of useful life had expired. Common to all was a need for the structure to be assessed so that a decision could be made as to whether it should be demolished and rebuilt anew, retained, adapted for another use, (upgraded if necessary), or simply restored to its original strength so that it could continue to serve its intended purpose. With the economy of the UK and many other countries continuing to be in recession it appears inevitable that this trend will continue towards maintaining existing construction at the expense of new building and it is against this background that this book has been written.

The author, of international renown, is now with Sir William Halcrow and Partners Ltd, and led the Materials Science Department of the Travers Morgan Consulting Group for many years, specializing in concrete technology. He writes with the benefit of continual, direct personal involvement in projects. At the design stage he seeks to influence the chosen solution so as to minimize the risk of the structure not fully serving its design life. However, much of his work is related to the existing building stock to which I have referred and this book is a complete guide to remedial structural work. It starts by whetting the reader's appetite for a fuller understanding of the complex nature of construction materials, before going on to describe the processes of deterioration.

Having determined the need for an investigation I fully support the author's proposals for designing the investigation, rather than allowing it to emerge as often seems to be the case. The extensive sections on testing procedures and the interpretation of results I believe will be helpful to many. The sections on repair and restoration techniques and protection not only bring the reader up to the forefront of current practice but also reveal a rapidly advancing technology.

I believe this book will be of extensive use both to experienced practitioners and to students for whom it will be an introduction to a subject of compelling interest. If reading it leads to a heightened awareness of possible pitfalls and how to avoid them by good design, then I know that Dr Kay will feel well rewarded.

K C WHITE, FEng
CHAIRMAN, TRAVERS MORGAN
CONSULTING GROUP

Disclaimer of warranty

Neither the author nor publisher makes any warranty, express or implied, that the calculation methods, procedures and programs included in this book are free from error. The application of these methods and procedures is at the user's own risk and the author and publisher disclaim all liability for damages, whether direct, incidental or consequential, arising from such application or from any other use of this book.

Acknowledgements

When I was younger, it was a cynical ambition of mine to be in the right place at the right time. This has happened on a few fortunate occasions but on an autumn evening in Newcastle, I found myself in the wrong place at the wrong time. It was at a reception to celebrate the first year in business of Travers Morgan's local office and both Professor Kew Kong and Mr K C White were present. Professor Kong enquired whether I had ever considered writing a book on concrete deterioration and repair. Mr White was enthusiastic about the idea and, as he was chairman of the company for which I was then working, there was no turning back. Mr White's encouragement continued throughout the writing of this book and he made numerous useful suggestions on improving the text.

Sincere thanks are due to Ann Andrews and Wendy Vick who prepared the manuscript and who dealt cheerfully with the many changes and revisions. They in turn wish to pass on their thanks to the inventor of the word processor.

Finally, I should like to dedicate this work to my family and thank them for their forbearance over the many lost evenings and weekends during the period when I was researching and writing this book. I think that I can promise them with confidence that it will never happen again.

We are grateful to the following for permission to reproduce copyright material:

The British Standards Institution for Table 5.3 (BS 8110: Part I: 1985); extracts from British Standards are reproduced with the permission of BSI. Complete copies can be obtained by post from BSI Sales, Linford Wood, Milton Keynes, MK14 6LE; the Controller of Her Majesty's Stationery Office for Fig. 8.2 (from *Materials for the Repair of Concrete Highway Structures, 1986*); The Institution of Structural Engineers for Figs. 5.14 and 5.15 (Doran, 1988), Fig. 5.12 (Choong Kog Y, 1989).

Whilst every effort has been made to trace the owners of copyright material, in a few cases this may have proved impossible and we take this opportunity to offer our apologies to any copyright holder whose rights we may have unwittingly infringed.

1 Introduction

It is the best of materials, it is the worst of materials. It is the material used with beauty and flare in the Sydney Opera House and also the material used for the construction of the ugliest tower blocks and the most dismal car-parks. It has mundane uses, it has spectacular uses. It has been used underground to line tunnels and in foundations; it has been used underwater for pipelines and harbours; it has been used high above ground in graceful arches and skyscrapers. Some early examples of its use are still in existence after a thousand years; some of its recent applications have deteriorated rapidly and have had to be demolished. Its name at one time became a synonym for permanence; the word is now used as an adjective of despair to describe the harsh environment produced by thoughtless over-use in our inner cities — the concrete jungle.

Concrete is one of the most versatile of construction materials. It presents many faces to the public and is perceived in many different ways in the modern world. Many of its applications are in combination with steel reinforcement and in this sense any examination of its performance is a tale of two materials.

Steel and concrete are complementary in several areas of their properties. Steel is strong in tension but, in the form of rod reinforcement, is not able to resist large compressive loads because of buckling instability. Concrete is weak in tension but strong in compression; its mass provides stability against buckling failure of contained reinforcement. Steel is ductile, concrete is brittle. Unprotected steel is subject to corrosion under normal atmospheric exposure; concrete is, superficially, chemically stable under these conditions. The alkalinity of concrete provides a passive environment where steel is less likely to corrode. The two materials possess similar coefficients of thermal expansion and so when they are combined they do not exert undue strains on one another when subjected to wide temperature extremes.

This match of properties should lead, in theory, to an almost ideal symbiosis. Over the past 50 or so years a huge number of concrete structures have been completed and the vast majority of these structures have given trouble-free service. However, the problems (sometimes severe) exhibited by a minority, serve to illustrate the unfortunate fact that the ideal is not always achieved. The object of this book is to show how these problems can arise, how they can be recognized from outward appearance and the results of fairly straightforward tests, how the deleterious effects can be minimized; and to give the techniques currently available for repair.

To assist in understanding the various deterioration mechanisms it is first necessary to learn something of the constituents and the manufacture of concrete and how these can affect its relevant properties.

Constituents

Concrete starts its life as a fluid mixture of manufactured and natural ingredients, some of which take part in a chemical reaction to produce the hardened stone-like product. The ingredients are cement and water which take part in the reaction, graded aggregate which acts as a filler and admixtures which act to modify some of the properties of the concrete in the fluid or hardened states. The cement may be Portland cement or a mixture of Portland cement and other hydraulic cement or pozzolanic material such as fly ash, blast-furnace slag or microsilica.

Manufacture of Portland Cement

The manufacture of Portland cement consists of heating together ingredients which contain the desired combination of calcium, silicon, aluminium and iron oxides until they fuse together to form a clinker. The clinker is ground down with the addition of gypsum to produce the familiar fine grey powder. Within this extremely broad framework several alternative processes are available. Historically, the raw materials were mixed together with water to form a slurry before introduction to the kiln. This is the so-called wet process which requires considerable consumption of energy to drive off the water before the fusion process begins. In more recent years the 'semi-wet', 'semi-dry' and 'dry' processes have been developed in the quest for energy efficiency.

The main raw materials for cement making are chalk or limestone and shale or clay. Chalk or limestone consists principally of calcium carbonate. In the kiln, carbon dioxide is driven off to leave calcium oxide. The shale or clay is the source of the oxides of silicon, aluminium and iron. In coal-fired kilns the coal ash can also be a useful source of these elements. Cement works in the United Kingdom have developed at places where there is a fortunate local availability of suitable materials. This proximity of suitable materials is not always the case in developing countries and there may be a need to import iron oxide or bauxite (an oxide of aluminium) from substantial distances.

As with most naturally occurring materials there can be considerable local variations in properties and constituents of the limestone, clay or marl, and such variations are undesirable in the raw material for a controlled industrial process. Inclusion of even minor quantities of certain impurities can lead to significant changes in the properties of the product. In limestone deposits, the impurities tend to occur in discrete bands or as veins which are visible in the quarry face. The effects can be minimized by careful planning of quarry operations.

The raw materials have to be broken down into a fine powder before introduction to the kiln. This is necessary so that individual particles react fully in the short period during which they are exposed to high temperature. The size reduction process is usually carried out in at least two stages after the initial breaking up, which takes place as an inevitable consequence of quarrying. The stockpiling which takes place between the first stage crushing and the second stage milling can be used as part of the blending process to produce ingredients of uniform composition. Crushing is usually carried out by compressive action or by impact, and the aim is to produce material typically in the size range 10−30 mm for storage and blending before milling. The actual size of material will depend on the type of mill to be used.

The milling process is also used to mix and dry the ingredients. Raw materials are fed to the mill in the required proportions with hot gas from the kilns for drying. In some cases the drying is carried out in a separate classifier. Most mills are horizontal rotating cylinders using steel balls, but some vertical spindle or ring roller mills have come into use. After the initial pass through the mill, the material is passed to a separator and coarse material is returned to the mill.

Final blending takes place in the storage silos which typically hold sufficient material for one day's production. The silos have aeration equipment in their base which is able to fluidize the stored material to assist final mixing before feeding to the kiln.

The Wet Process

In the wet process the ground ingredients are mixed together with water to form a slurry with typically 30 to 40 per cent moisture content before introduction to the kiln. The kiln consists of a long inclined cylinder lined with refractory material. Kilns for the wet process may be up to 6 m in diameter and 200 m in length. The kiln rotates as the raw material is fed into the top end and fuel is burned near the bottom end to produce a counter-stream of hot gas. As the raw material falls down the kiln it first encounters a region where chains hang across the kiln. In this region, the materials are dried and formed into nodules. Further down the kiln, carbon dioxide is driven off and the materials fuse together to form clinker as a temperature of over 1400 °C is reached. At the lower end of the kiln, the clinker is allowed to cool before emerging at a temperature around 1100 or 1200 °C.

The drying process in the kiln is wasteful of energy and modern plants tend to be designed to use dry processed ingredients as described above. Semi-wet and semi-dry processes have also been developed to reduce energy consumption.

Semi-wet Process

The raw materials are prepared as a slurry as for the wet process and fed into a filter press where the moisture content is reduced to approximately 20 per cent. The filter cake is broken up before being fed to the kiln, possibly via a preheater.

Semi-dry Process

The semi-dry process is also known as the 'Lepol' process and involves dry preparation of the ingredients as described above before they are formed into pellets with the addition of approximately 12 per cent water. The pellets are dried and preheated in a moving grate, using kiln exhaust gases, before passing to the kiln proper.

Dry Process

The most energy-efficient cement-making process uses dry ingredients (0.5 per cent moisture content). The dry ingredients are fed through a preheating system of cyclones in which the kiln exhaust gases are used to raise their temperature. In some systems, additional fuel is used in the heaters to drive off the carbon dioxide from the chalk or limestone. Material enters the kiln at a temperature of between 750 °C and 950 °C which means that dry-process kilns can be much shorter than those used in the wet process. The essential differences between the various cement making processes are illustrated in Fig. 1.1.

Cooling and Grinding

When the clinker emerges from the kiln it is still at a high temperature. Clinker is cooled before grinding and air flow is used to accomplish this in a number of alternative processes. The clinker may be passed into a rotary drier which is cylindrical in shape, or may be dried in smaller cylinders arranged around the outside of the main kiln body. A third

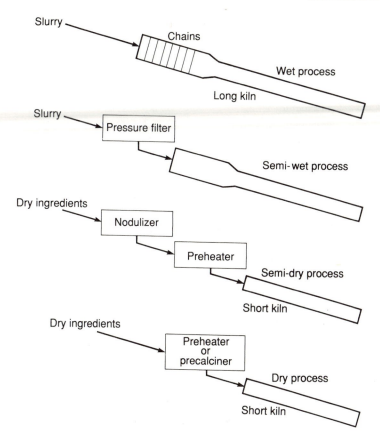

Fig. 1.1 Differences between main cement-making processes

alternative is to spread out the clinker on a moving grate. To increase energy efficiency, the warm air from the cooler may be passed into the kiln as part of the air supply for combustion.

Grinding is carried out in a ball mill divided into a number of chambers by slotted diaphragms which allow the cement to pass through but which retain the steel balls. The chambers permit successively smaller balls to be used as the cement powder becomes finer. Gypsum is added to the clinker before grinding as this acts as a set regulator. The heat generated by grinding is dissipated by spraying with cold water. This is done either onto the outside of the mill or by water spray injection at the delivery end. In the latter case, the amount of water has to be carefully regulated as only sufficient water to produce cooling by evaporation must be injected.

After grinding the cement passes to bulk storage silos for onward delivery or bagging. A full account of the modern cement-making processes has been given by Kerton and Murray.[1.1]

Portland Cement Composition

The basic raw materials of cement, calcium carbonate, iron oxide, alumina and silica react together in the kiln to produce the four major constituents of cement:

Tricalcium silicate	$3CaO \cdot SiO_2$	(C_3S)
Dicalcium silicate	$2CaO \cdot SiO_2$	(C_2S)
Tricalcium aluminate	$4CaO \cdot Al_2O_3$	(C_3A)
Tetracalcium aluminoferrite	$4CaO \cdot Al_2O_3 \cdot Fe_2O_3$	(C_4AF)

Cement chemists use a shorter method of annotation whereby each of the main oxides is referenced by one letter. Hence CaO is represented by C, SiO_2 by S, etc., and water is represented by H. Using this notation the major cement constituents have the formulae shown in brackets above.

Cement also contains minor quantities of other materials such as the oxides of the alkali metals, sodium and potassium. These have important consequences as they contribute to the alkalinity of concrete, which is important for the passivity of embedded steel, and they may also react with certain aggregates.

Hydration Reactions

When cement is brought into contact with water the constituents hydrate at different rates and make contributions to the overall strength of the concrete at different times. Tricalcium silicate assists in the initial set and continues hydrating for the first few days. The hydration reaction can be represented as follows:

$$2C_3S + 7H \rightarrow C_3S_2H_4 + 3CH$$

This reaction makes a significant contribution to the early strength. Dicalcium silicate hydrates at a much slower rate with similar products to those of tricalcium silicate but much less calcium hydroxide (portlandite) is produced:

$$2C_2S + 5H \rightarrow C_3S_2H_4 + CH$$

This reaction proceeds for an extended period and is thought to be the main contributor to strength beyond 28 days.

Tricalcium aluminate undergoes a very rapid reaction in the presence of water which can result in flash set. Gypsum is added to cement to control this reaction. In the presence of sulphate, tricalcium aluminate reacts with water to form a calcium sulphoaluminate known as ettringite:

$$C_3A + 3CSH_2 + 26H \rightarrow C_6AS_3H_{32}$$

If most of the sulphate is consumed in the reaction the ettringite becomes unstable and converts to a different sulphoaluminate hydrate:

$$C_6AS_3H_{32} + 2C_3A + 4H \rightarrow 3C_4ASH_{12}$$

Tricalcium aluminate gains most of its strength in one day and little thereafter. Tetracalcium aluminoferrite also reacts fairly rapidly with water but is not thought to add much to the strength of the concrete.

In summary, C_3A has its main effect in the rheology of the fresh mix and the setting; C_3S gains most of its strength in the first 7 days and little increase occurs at later ages; C_2S contributes little to early strength but continues gaining in strength after 28 days; C_4AF makes little contribution to strength.

Changes in Cement

The composition of cement has not remained constant over the years and this has to be kept in mind when investigating the condition of old structures and assessing their strength. As the manufacturing processes have changed the properties of the product have changed also. Cement manufacturers have responded to an apparent demand from users for stronger cements and cements which gain strength quicker. Stronger cements mean that less cement has to be used in a mix to produce a concrete of given strength and, as cement is the most expensive ingredient, this can lead to significant economies. Cement which gains strength more quickly means that formwork can be stripped earlier, leading to shorter construction periods; in precast works, moulds can be stripped earlier and put back into use. Both of these also result in cheaper construction.

The Concrete Society[1.2] has carried out an investigation of the changes in cement properties and their effects in the United Kingdom. They have found that in the period between 1914 and 1984 there was a significant change in the proportions of C_3S and C_2S in cements in the United Kingdom and also in the United States and Denmark. In the years between 1914 and 1922, the range of C_3S content in United Kingdom cements was from 15−48 per cent. By 1984 this had risen to 54−63 per cent. In the same period, there had been a lowering of the range of C_2S from 15−26 per cent in 1914−22 to 8−27 per cent in 1984.

The investigation found also that there had been a significant rise in strength particularly at early ages. Results were available from 1:6 concretes made with a water/cement ratio of 0.6 and tested at various ages. In the period 1950−54 the mean strength at 28 days had been $34 \, \text{N mm}^{-2}$. In 1984 the 28 day strength had increased to $45 \, \text{N mm}^{-2}$.

Over the same period, the ratio (7 day) : (28 day) strength had risen from 0.65 to 0.71.

The effects of these changes in cement properties were examined in several key areas which are of interest in the context of structural assessment and rehabilitation. It was found that concrete of a given workability and strength could now be made with a lower cement content and a higher water/cement ratio. As an example, a $30\,N\,mm^{-2}$ concrete with medium workability would have required $440\,kg/m^3$ of cement in the 1950s and a water/cement ratio of 0.43. The equivalent mix in the 1980s required $350\,kg/m^3$ of cement and a water/cement ratio of 0.54. The Concrete Society report pointed out that these changes could have a marked effect on durability as the lower cement content and higher water/cement ratio would result in more permeable concretes. This greater permeability leaves the concrete more susceptible to attack by frost, chloride penetration and sulphate attack. In addition, the more rapid reaction due to the increased C_3S contents could lead to increased heat evolution and consequent thermal cracking.

Structure of Cement Paste

More water is always added to concrete than that merely required to take part in the hydration reaction with the cement. The extra water is required for workability as the concrete is mixed and placed in position.

In water, the cement grains tend to join together in small groups leaving water filled gaps between each group and the next. As each cement grain hydrates, a halo of cement gel forms around it and calcium hydroxide crystals grow in some of the gaps. The haloes reduce the contact between the cement grains and water, but the reaction proceeds and gradually the space between the grains begins to fill with gel. Pores within the gel trap some of the water and there may be relics of unhydrated cement at the locations of the centres of the original grains.

The cement gel is not able to fill completely the space between the groups of grains. As the gel grows and the cement paste sets, water is trapped in capillary passages. Hydration continues and as a consequence the volume of water decreases as it is used up in the reaction until there is insufficient to fill the capillaries, and so pockets of water vapour form. The process of structure formation is shown diagrammatically in Fig. 1.2.

At first the capillaries are interconnected and this remains the case if the concrete has too high an original water content or if hydration is not permitted to continue over a sufficiently long period. It has been shown[1.4] that, so long as sufficient water is provided, the interconnected pores in concretes with a low water/cement ratio will eventually become blocked. However, as shown in Fig. 1.3, as the water/cement ratio

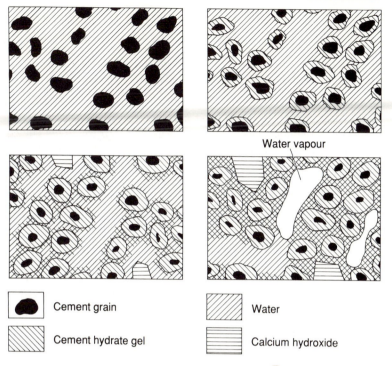

Fig. 1.2 Development of capillaries in cement paste structure (after Bennett[1.3])

Water vapour

Cement grain	Water
Cement hydrate gel	Calcium hydroxide

increases, the time for which water curing must be provided increases substantially until for values in excess of 0.7 it is not possible to block the pores.

The presence or absence of a system of interconnected pores within the cement paste is of prime importance in considerations of concrete durability as they control permeability. In general, permeable concretes

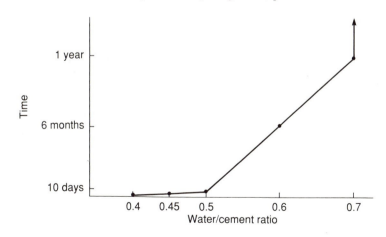

Fig. 1.3 Length of curing required to block capillary pores for concretes of different water/cement ratios (after Neville[1.4])

are much less durable as the pore structure allows air, water and any dissolved salts to penetrate deep into the body of the concrete.

Other Cementing Materials

A number of materials with either hydraulic cementing or pozzolanic properties are available as by-products from industrial processes. They have been used in concrete for many years, but their use in the United Kingdom has become more widespread since the publication in 1985 of BS 8110[1.5] which permits the use of pulverized fuel ash and ground granulated blast-furnace slag. Microsilica has also been used in concrete in the United Kingdom, but its use has been more extensive in some continental countries, particularly Norway where its use is permitted by standards.[1.6]

Pulverized Fuel Ash

Many modern power stations burn pulverized bituminous coal as their fuel. In the United Kingdom, the coal used as fuel in many power stations contains a mixture of carbonaceous matter with a variety of minerals in shales and clays along with sulphides and carbonates. It is first pulverized to a size such that most passes a 75 μm sieve before being blown into the furnace. The fuel is ignited and burns at a temperature of around 1600 °C which is sufficient to melt most of the minerals it contains. Some of the residual ash falls to the bottom of the furnace, but the finer particles are carried away with the exhaust gases.

Solid particles in the flue gases are collected either in a combination of cyclones and electrostatic precipitators or in precipitators alone. The residue is known as pulverized fuel ash (PFA) or fly ash. It is predominantly the finer particles that are obtained from the precipitators and it is this fraction which is used in concrete.

As the clay minerals in the coal pass through the ignition process in the furnace, they are converted into glassy spheres predominantly of aluminosilicate but also containing calcium, iron, magnesium, potassium and sodium. Some of the carbon is unburnt and particles of carbon will also be present in the ash. The properties of the ash depend upon the fuel being burnt and the conditions within the furnace. For instance, there is likely to be more carbon present in the ash produced when the furnace is starting up or shutting down. Power stations which are used to provide the base load of power demand therefore produce a more consistent ash than those which operate only at peak periods.

The material produced in the electrostatic precipitators is extremely fine and does not require grinding. However, it will usually pass through classification and quality control procedures before being released for use in cement.

Properties of PFA

The glassy aluminosilicates are present in PFA as spherical particles. They do not undergo a cementing reaction with water alone but the material is pozzolanic, i.e. it reacts with calcium hydroxide in the presence of water, forming calcium silicate and aluminate hydrates similar to those produced when Portland cement reacts with water. Calcium hydroxide is produced in the hydration reaction of cement and this is the normal source used with PFA. Cement and PFA may be blended together and sold as Portland PFA cement but the more usual practice in the United Kingdom is to add them as separate ingredients at the mixer.

Clearly the pozzolanic reaction requires the initial production of calcium hydroxide by cement hydration and so PFA contributes little to the early strength of concrete mixes. However, as time goes on the reaction with PFA continues and can cause appreciable gains in long term strength. Another consequence of the slow start of the reaction is that heat of hydration is not liberated so quickly and the peak temperature is reduced.

Residual carbon in PFA often occurs as highly porous particles. These particles tend to absorb some admixtures and reduce their effectiveness in concretes containing PFA. This effect has been noted, particularly in relation to air entraining admixtures, and there may be a resulting need to increase dosage when PFA is used.

Blast-furnace Slag

Iron is produced in blast-furnaces which are charged with the raw materials of iron ore, coke as a source of carbon, and limestone which acts as a flux. In the smelting process iron is produced in molten form and a slag forms on its surface. The slag results from the fusion of limestone with ash from the coke, and aluminates and silicates from the ore, and the compounds which the slag contains are similar to those in Portland cement.

Blast-furnace operation is continuous and the ingredients added to the furnace are carefully controlled to obtain a consistent product. This being the case, the slag produced at a particular works also has consistent composition unless the process is changed. The slag is lighter than the molten iron and can be run off from the top of the furnace. To produce a material suitable for use as cement, it is necessary to cool it quickly so that it solidifies in a glassy state. This is achieved by pouring a stream of molten slag through a fine spray of water or cold air and water in a granulating chamber. The slag forms into glassy pellets of consistent particle size which fall to the bottom of the chamber and which are collected in water-filled tanks. The material is then dried and ground in mills similar to those used for grinding cement clinker, to produce ground granulated blast-furnace slag (GGBS). As for PFA the GGBS may be interground with Portland cement or added as a separate ingredient at the mixer.

Properties of GGBS

The glass produced by the rapid cooling of molten slag is a disordered network of oxides of calcium, silicon, aluminium and magnesium. In order for it to react, it requires activators to break down the glassy structure. Suitable activators are found in the alkalis and sulphates, which are produced in the hydration reaction of Portland cement. Once the reaction is under way calcium silicate and calcium aluminate hydrates are produced directly by the reaction with water. Calcium hydroxide is also produced and the reaction is not pozzolanic since it does not depend upon combination of the slag with calcium hydroxide. As with PFA, the rate of strength gain and peak hydration temperatures of mixes containing GGBS are lower than the equivalent ordinary Portland cement mixes.

Microsilica

Microsilica, or condensed silica fume, is a by-product of the electric arc furnace reduction of quartz into silicon and ferrosilicon alloys used in the electronics industry. The process is usually carried out in countries where electric power is cheap and some 2 000 000 tonnes are produced annually throughout the world.

Microsilica is produced as an extremely fine powder (100 times finer than cement) consisting of spherical particles with a silica content in excess of 85 per cent. It is difficult to use in this form because of its low bulk density, and is usually processed in one of several ways. Densified microsilica consists of particles loosely cemented together in agglomerations which can be broken down if used in high shear mixers. The material can also be supplied as a 50:50 (by weight) slurry mixed with water, which is a convenient form for use in concrete mixes of all kinds. A third alternative is to form the microsilica into small pellets of 0.5−1.0 mm in diameter. The material in this form can be interground with ordinary Portland cement to produce a blend.

Properties of Microsilica

The action of microsilica in concrete is pozzolanic, i.e. it reacts with the calcium hydroxide which is produced in the hydration reaction of Portland cement. In this case, as microsilica consists predominantly of silica, the main reaction product is a calcium silicate hydrate. However, the presence of microsilica in the mix apparently has other effects.[1.7] The fine particles of microsilica have a nucleation effect which assists the hydration of tricalcium silicate in the cement.

The strength gain of microsilica cement concretes may be slower than that of equivalent Portland cement mixes at early ages, but the development of strength progresses much beyond 28 days. Microsilica has been used in concrete of very high strength (up to 130 N mm^{-2}). The effect on heat of hydration is not yet clear.

Aggregates

Aggregates are used in concrete as inert fillers to bulk out the volume, and they are not intended to take part in the hydration reaction. However, their properties, particularly grading and particle shape, can have marked effects on the plastic behaviour of concrete, which have consequences for its long term performance in the hardened state. Some mineral types present in certain aggregates have been found to take part in a reaction with alkalis from the cement, resulting in disruptive expansive forces.

Excavation

Most concrete aggregates are stone, sand or gravel from naturally occurring deposits, but some manufactured aggregates of clinker or slag are sometimes used. Special lightweight aggregates are also manufactured from clay or pulverized fuel ash.

Rock in massive deposits is usually won by quarrying. After the overburden has been stripped, the rock is broken up by blasting or ripping with mechanical plant to produce material of suitable size for processing. Blasting is a carefully calculated and controlled operation if it is to be carried out economically and to produce the correct degree of fragmentation. This is important because producing material of the correct size means that it can be handled more easily, causes less wear on machinery and is less likely to cause blockages in the crusher which reduce throughput.

Many sand and gravel deposits occur close to the surface and are won by a variety of digging mechanisms. Front-end loaders, back-hoes, face-shovels and drag-lines are all used. Water jets are also used in locations where there are ample supplies of water. As the supplies of land-based aggregates have become exhausted in industrialized countries and planning applications have become more difficult, there has been an increase of interest in the marine gravels on the continental shelf. These can be exploited using grabs, ladder dredgers or cutter suction dredgers.

Crushing

Crushing is designed to reduce rock to a particular size range while minimizing the proportion of finer material produced. Machines have to be of robust construction and they use relatively large amounts of energy. Most crushers work by trapping the rock between two metal surfaces, which are moved relative to one another.

In jaw crushers the material is fed under gravity between two almost vertical plane faces, one of which is fixed and the other of which moves on a pivot. Gyratory crushers consist of an upward pointing cone of shallow angle inside a truncated downward pointing cone. The inner cone rotates eccentrically to the outer cone with the result that at any particular point the gap between the two cones opens and closes continually. Material is fed in from the top and is crushed as it is trapped between the inner and outer cones.

A similar arrangement is used in a cone crusher except that both cones are upward pointing and the crushing faces are almost parallel. The material spreads out as it works its way down to the discharge point and therefore blockage is less likely.

Impactors work on a different principle; they consist of horizontally mounted wheels with paddle-like beaters on their circumference. The wheel rotates at high speed while a stream of material is fed onto it. The kinetic energy from the beaters causes some fragments to shatter, but all of the material is flung against an impact plate where a further reduction in size occurs. One advantage of this type of equipment is that it tends to produce aggregate which is more cubical in shape. The mechanical principles of the different forms of crusher are illustrated in Fig. 1.4.

Grading

Grading is the process of dividing the crushed, or in the case of many gravel winning operations as-dug, material into fractions of different sizes so that it can be used in a controlled way in the production of concrete. The process is carried out in most cases by use of wire mesh, parallel bars or metal plates with holes of suitable size. This is known as a screening operation. At the finer end of the scale, the particle settlement

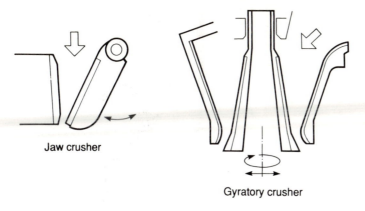

Fig. 1.4 Comparison of crushing methods (after Collis and Fox[1.8])

Jaw crusher

Gyratory crusher

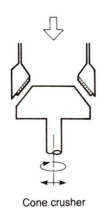

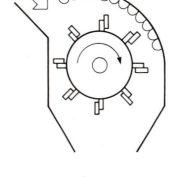

Cone crusher

Impactor

velocity in water is used to separate out different sizes and the process is known as classifying.

Most screening is carried out on inclined vibrating screens which use gravity to assist the movement of material. It is usual to have the screens arranged so that the coarsest material is removed first working downwards to progressively smaller sizes.

Classifying operations use fairly straightforward equipment with few moving parts. One method is to use a water-filled tank to which the material is delivered in slurry form through a vertical tube in the base. The fine material is carried upwards with the water current and is discharged over the side of the tank. Meanwhile, the coarser fractions settle back to the base of the tank from which they can be removed by Archimedes screw or bucket elevators.

An alternative is to use a long narrow tank to which the slurry is introduced at one end. The finer material is carried all along the tank and is discharged over a weir at the remote end. The base of the tank

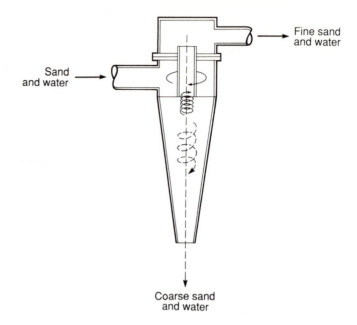

Fig. 1.5 Cyclone for sand separation (after Collis and Fox[1.8])

Sand and water

Fine sand and water

Coarse sand and water

is divided into a number of separate compartments which collect coarse material near the inlet and progressively finer fractions toward the outlet.

Cyclones are also extensively used to reduce the amount of fine material in sand. As shown in Fig. 1.5 they consist of a hollow cone with the point downwards and vertical discharge at top and bottom mounted on the axis of the cone. The slurry is introduced tangentially near the top of the cone-shaped chamber and creates a vortex which forces the coarser particles outwards by the action of the centrifugal force. At the outer wall there is a zone in which the flow is downwards. This carries the coarse particles to the discharge point at the base. The downward flow at the wall is compensated by an area of upward flow near the centre of the vortex. Here the finer particles are carried upwards to the discharge point at the top of the unit.

Washing

Many sand and gravel deposits contain clay and silt which must be removed. Marine deposits may have to be washed to remove chlorides which could cause corrosion of reinforcement. Many scrubbers used for this operation consist of long hollow cylinders partly filled with water and gravel. The cylinder has longitudinal ribs fixed to its inner surface and, as the cylinder is rotated, the gravel is lifted up and eventually falls back under the action of gravity. The tumbling action breaks up any clay lumps to form a slurry with the water. Gravel and water are fed into one end of the cylinder and clean gravel and slurry is delivered from the other end in a continuous operation.

The occurrence, field investigation, extraction, processing and use of aggregates of all types are described comprehensively in a report by a working party of the Engineering Group of the Geological Society.[1.8]

Properties

The properties of aggregate for use in concrete are covered by BS 882[1.9] which includes fine, coarse and all-in material. In this context fine aggregate is material that mainly passes a 5 mm test sieve. The properties covered by BS 882 include particle shape, shell content, mechanical properties and grading. An appendix gives guidance on chloride content.

Shape is specified according to the flakiness index. The limits depend upon whether the aggregate is crushed or uncrushed, and the grade of concrete in which the material is to be used. For uncrushed gravel in concrete with a strength of 20 N mm^{-2} or 35 N mm^{-2}, the limit is 50. For crushed rock or crushed gravel in concrete of strength 20 N mm^{-2} or 35 N mm^{-2} the limit is 35, as it is for all aggregates in concrete above 35 N mm^{-2}. There are no requirements for lower strength concrete.

The specified limits on shell content are 8 per cent for fractions above 10 mm and 20 per cent for the size range $5-10$ mm. There are no requirements on shell content for aggregates finer than 5 mm.

The mechanical strength of aggregate is judged on the basis of the 10 per cent fines value. In essence, this is the crushing force which must be applied to a sample of the aggregate in order to produce 10 per cent (by mass) of fines, and BS 882 gives different limits according to the situation in which the concrete is to be used. For general application, the value must exceed 50 kN; for the wearing surfaces of pavements, a minimum value of 100 kN is specified while for heavy duty floor finishes the corresponding figure is 150 kN.

Particle size is controlled by reference to grading envelopes for single-size coarse material and three grading zones for fine aggregate. Additional grading envelopes are specified for all-in aggregate. Limitations are placed on the maximum clay, silt and dust content ranging from 1 per cent for gravel to 15 per cent for crushed rock fines.

Maximum chloride content limitations are dependent upon the end use of the concrete and the type of cement to be used. Prestressed concrete or steam-cured concrete is limited to 0.02 per cent chloride ion by mass of aggregate. Concrete made with sulphate resisting cement is restricted to 0.04 per cent, whilst ordinary Portland cement concrete containing embedded metal is limited to 0.08 per cent.

Problems with Aggregates

A number of problems may arise because of the use of aggregates with imperfect properties, which are mainly related to particle shape and grading. Poor grading may lead to excessive bleeding and segregation. Bleeding is upward movement of water in the concrete mix which may lead to the formation of a much weaker layer on the surface of the hardened concrete, which is then susceptible to attack. The bleed channels provide a passageway by which aggressive agencies can find their way into the body of the hardened concrete. The upward bleeding of water may be accompanied by downward movement of other ingredients of the mix, resulting in cracking. Aggregates which contain an excess of fine material have a high water demand resulting in a less durable concrete.

Aggregates with poor shape are likely to result in harsh mixes which will require additional water to improve workability and which may be prone to segregation. Aggregates of low mechanical strength may break up in the mixing process, raising the fines content and the water demand of mixes. If the particles survive the mixer, the aggregate may not be strong enough to withstand the exigencies of service and may break up causing pop-outs on the surface.

Some aggregates have been found to cause excessive shrinkage, resulting in cracking of members. Sea-dredged aggregates, beach sands and aggregates in desert countries may contain concentrations of chlorides which are sufficiently high to cause corrosion of reinforcement in concrete. Aggregates from some locations have reacted with alkalis from the cement, leading to internal expansive forces and cracking. Many of these topics are discussed in more detail in Chapter 2.

Water

Water is usually the least troublesome of the ingredients of concrete in most of the locations where construction takes place. There is usually a ready mains supply of potable water which is perfectly suitable for concrete production. However, this happy situation is not always the case in developing countries where the contractor may have to rely on supplies from local wells. The concern is that the water may contain dissolved salts in sufficient concentrations to affect the setting properties of concrete or to cause corrosion of the reinforcement. The salt content can vary at different times of the year and reliance should not be placed on a single analysis at the beginning of the production period.

In the past, sea water has been used either for mixing or curing concrete. For unreinforced concrete this may have been of little consequence except for the accelerating effect, but there are many cases where corrosion has shortened the life of reinforced members.

Admixtures

Admixtures are materials added to concrete during the mixing process to modify its properties. They are usually added in relatively small quantities and may be used to modify the properties during the plastic state or in the hardened state. Examples of the former case are plasticizers and retarders, while examples of the latter application are given by air-entraining agents and integral waterproofers. There are many different types of admixtures,[1.10] but they can be divided into five main classifications as described in the following sections.

Accelerating

These admixtures are defined in BS 5075[1.11] as materials that increase the initial rate of reaction between cement and water, and thereby accelerate the setting and early strength development of concrete. Historically, calcium chloride was the most common and widely used accelerator both in the United Kingdom and the United States, but its use has been curtailed in reinforced concrete because of its effect on reinforcement corrosion. Its use is banned in the United Kingdom for any concrete containing embedded metal, though it is perfectly acceptable for use in concrete which contains no reinforcement.

Other accelerating admixtures are now available which do not contain calcium chloride. They may be based on calcium formate, calcium nitrate or lithium oxalate. The exact mechanism by which these materials increase the rate of hydration is not known. In the case of calcium chloride, it is thought that chloride ions diffuse into the C_3S and C_2S grains with an accompanying outward diffusion of hydroxyl ions. The rate of deposition of calcium hydroxide and decomposition of calcium silicates is increased.[1.11]

The main use is to allow early removal of formwork and also to permit concreting to continue during cold weather. However, the concrete still needs to be protected against frost in the early stages.

Retarding

The BS 5075 definition for these admixtures is that they are materials which decrease the initial reaction between cement and water, and thereby retard the setting of concrete. The main active components are lignosulphonates, or their derivatives, and hydroxycarboxylic acids, and their salts. Lignosulphonates are naturally occurring materials found in timber and it is their sugar content which leads to retardation.

The mechanism by which retarders work is thought to involve their absorption by C_3A and the formation of a film around individual cement grains which reduces their contact with water. The protective film eventually breaks down and the hydration reaction proceeds.

Retarders have been used to extend the time during which the concrete can be vibrated and hence assist in the concreting of large pours and in placing of concrete during hot weather.

Water Reducing/Plasticizing

As can be deduced from the dual classification, these admixtures may be used in two ways. They can be used to reduce the amount of water that has to be added to achieve a given workability or they can be used to increase the workability of concrete with a given water content. When used to reduce water, the resulting concrete is consequently stronger for a particular cement content, and hence water-reducing admixtures can be used to reduce the amount of cement required to achieve a given strength grade, with a consequent saving in cost.

The materials used in the production of these admixtures are similar to those used in retarding agents. Lignosulphonates may undergo additional refining to remove sugars which would otherwise lead to retardation. In addition to the use of lignosulphonate and hydroxycarboxylic acid salts in normal plasticizers, some derivatives of formaldehyde, such as melamine formaldehyde, are used as superplasticizers. They are capable of producing flowing or self-levelling concrete, or alternatively mixes of normal workability but with extremely low water content.

The materials used as water-reducing agents are surface-active agents. On the molecular scale, they consist of a water soluble head and a hydrocarbon tail. When these materials are placed in water they migrate to a free surface and when at the surface, they orient themselves so that their heads, which attain a negative charge, are just within the water and their tails stick out into the air.

In untreated mixes, the individual cement grains tend to congregate together in small groups. When a water-reducing admixture is introduced, the molecular tails are adsorbed onto the surface of cement grains leaving the negatively charged head protruding into the surrounding fluid. This leaves each grain, with the molecules of admixture attached, bearing a net negative charge as shown in Fig. 1.6 and the result is that grains tend to repel each other and the groups break up. The net effect is to increase the fluidity of the mix, and to expose a greater surface area of the cement grains to potential reaction with water.

Water reducing/plasticizing admixture

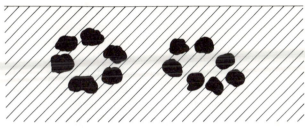

Figure 1.6 Break up of groups of cement particles by surface-active behaviour of water-reducing plasticizing admixtures

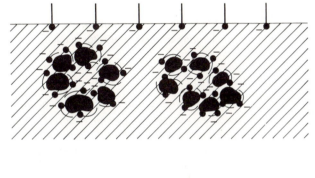

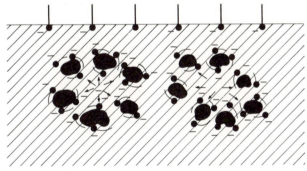

Similarly, any air bubbles that are present in the mix will develop a negative charge because of the way the admixture molecules align themselves. This means they are also repelled by the cement particles and, as a consequence, larger pockets of entrapped air are more readily removed. On the other hand, there is a tendency for a network of very stable small air bubbles to form in the mix and the same materials can be used to entrain air. If only a plasticizing action is required, the admixtures are formulated to contain an air-detraining agent, such as tributyl phosphate.

Water reducing/plasticizing agents have been put to a number of different uses:

1. Concrete can be made with greater workability without the use of additional water and so the strength is not reduced.
2. The water content can be reduced while maintaining workability, so a stronger concrete can be made with a given cement content.
3. The cement content can be lowered while retaining the same water/cement ratio and workability, to achieve a given strength.
4. Highly fluid concretes which require little vibration can be produced for use in 'self-levelling' situations and repairs.
5. Marginal harsh and poorly graded aggregates can be used to produce reasonable mixes.

Water-reducing admixtures are also available blended with other materials to produce accelerating or retarding water-reducing admixtures.

Air Entraining

As stated above, the materials used for air entrainment are organic surfactants which lower the surface tension of water. They facilitate bubble formation and the bubbles are uniformly dispersed because the negative charges on the water-soluble ends of the admixture molecules repel one another as shown in Fig.1.7.

Materials used for air entrainment are based on natural wood resins, some animal and vegetable oils and fats, such as tallow and olive oil, and lignosulphonates. In the United States, three grades of air-entraining cement are manufactured. The basic purpose of air entrainment is to improve durability under freeze/thaw conditions, though it may also

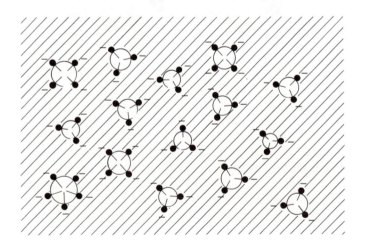

Fig. 1.7 Action of surface-active agents in air entrainment (air bubbles are indicated by circular symbols)

reduce the tendency of mixes to bleed and improve workability as the bubbles act like fine aggregate particles. Air entrainment has one important side-effect in that it reduces the strength of the hardened concrete, and this has to be allowed for in mix design. Air entrainment is used particularly on concrete highways and other exposed slabs.

Bubbles produced by air entrainment are spherical in shape, 0.05−1.25 mm in diameter and spaced less than 1 mm apart. The air contents employed in concrete are typically of the order of 5 per cent for 20 mm aggregate. It is thought that the freeze/thaw resistant properties can be attributed to the bubbles breaking up the network of capillaries in the concrete. The bubbles are too big to be filled with water by capillary effects and hence act as an expansion chamber if freezing occurs.

Waterproofers

Integral waterproofers are less widely used than the other categories of admixture. They may act by blocking the pores in concrete, or by lining the pores and rendering them hydrophobic or water repellent. Some admixtures use a combination of the two mechanisms. The pore blockers will tend to resist the flow of water at moderate pressure gradients, but pore liners will only work under low pressures and will not prevent the passage of water vapour.

Some materials that are used as pore blockers are bentonite, bitumen emulsion, colloidal silica and finely divided limestone. Stearates of ammonia and sodium, petroleum jelly and naphthalene are used as pore liners.

Dispensing

Admixtures are dispensed as liquids or powders. In the case of powders they are added to the other dry ingredients just before mixing. Most liquid admixtures are added to the mixing water, but some superplasticizers may be added to the mix at site because the high workability is only obtained for a short period of time and otherwise the full benefits would be lost.

In all cases it is essential that the materials are dispensed in the correct quantity or their effects will be diminished in the case of underdosing or exaggerated in the case of overdosing.

Admixtures/Durability

Use of admixtures can effect the long term performance of concrete in many ways. The dispersion of cement grains caused by plasticizing agents leads to more efficient hydration and consequently a less permeable concrete. Increased workability should mean that the concrete is more easy to compact and a denser product should result. If plasticizers are used to reduce the water content while maintaining the same cement content, a less permeable concrete should again be a consequence. However, if plasticizers are used to reduce cement content to achieve a particular strength, durability may suffer. As has been stated previously, the use of calcium chloride as an accelerator can lead to reinforcement corrosion. Overdosage of air-entraining admixtures can lead to reductions in strength which could lead to structural problems; underdosage can lead insufficient entrainment of air and susceptibility to frost damage.

Overdosage with plasticizers may lead to segregation or bleeding. The extended period of workability resulting from over-retarding may make finishing difficult but also means that the concrete is susceptible to drying winds for longer periods, and consequently there may be greater risk of plastic cracking.

References

1.1 Kerton C P, Murray R J 1983 Portland cement production. In Barnes P (ed) *Structure and Performance of Cements* Applied Science Publishers, London and New York, pp 205−36

1.2 Concrete Society 1987 *Changes in the Properties of Ordinary Portland Cement and Their Effects on Concrete* Technical Report No 29, The Concrete Society, Slough

1.3 Bennett K 1977 Air entraining admixtures for concrete. In Rixom M R (ed) *Concrete Admixtures, Use and Applications* The Construction Press, Lancaster, 37−47

1.4 Powers T C, Copeland L E, Mann H M 1959 Capillary continuity or discontinuity in cement pastes. *Journal of the Portland Cement Association* **1(2):** 38−48

1.5 British Standards Institutiom 1985 *The Structural Use of Concrete* BS 8110, The British Standards Institution, London

1.6 Norwegian Standards 1978 *Concrete Structures: Materials, Construction and Control* NS 3474, Norwegian Standards Institute, Oslo

1.7 Concrete Society 1991 *Microsilica in Concrete* Technical Report No 41, The Concrete Society, Slough

1.8 Collis L, Fox R A 1985 *Aggregates: Sand, Gravel and Crushed Rock Aggregates for Construction Purposes* Engineering Geology Special Production No 1, The Geological Society, London

1.9 British Standards Institution 1983 *Aggregates from Natural Sources for Concrete* BS 882, The British Standards Institution, London

1.10 Hewlett P C 1977 An introduction to and a classification of cement admixtures. In Rixon M R (ed) *Concrete Admixture Use and Applications* The Construction Press, Lancaster, pp 9−22

1.11 British Standards Institution 1982 *Concrete Admixtures, Specification for Accelerating Admixtures, Retarding Admixtures and Water Reducing Admixtures* BS 5075, The British Standards Institution, London

1.12 Diamond S 1980 Accelerating admixtures. *Proceedings of the International Congress on Admixtures* The Construction Press, Lancaster, pp 17−31

2 Deterioration Processes

It is important to gain an understanding of the basic causes and mechanisms of the various forms of deterioration which may attack concrete. Armed with this understanding, it is possible to undertake realistic assessments of the current condition of concrete structures, and to design and implement the appropriate remedial or refurbishment techniques.

Although deterioration of concrete structures is usually a medium to long term process, the onset of deterioration and its rate may be influenced by the presence of defects which have their origin at the time of construction, or in the very early stages of the life of the structure. These defects permit the atmosphere and other environmental agencies to penetrate the surface of the concrete and to take part in the chemical and physical processes which give rise to deterioration. This chapter will first discuss construction defects and early age behaviour and, in later sections, the most common processes which attack concrete will be examined.

Construction and Design Defects

Construction and design faults account for a surprisingly high proportion of defects in structures. An analysis[2.1] of 10 000 defects reported to French insurance companies during the period 1968—78 has shown that 88 per cent were associated with errors in design or construction. The analysis included all aspects of building work, including roofing and services and was not specific to concrete structures. However, bearing in mind the well-publicized cases, such as the use of high alumina cement, calcium chloride as an admixture and the many cases of spalled concrete due to failure to achieve adequate cover, it is considered probable that a significant proportion of the problems associated with concrete structures can be traced back to design, in its widest sense, or to construction defects.[2.2]

Design can affect the performance of concrete in a number of ways.

Provision of falls and drainage on slabs and other horizontal members will reduce the time during which the surface of the concrete is in contact with water and reduce the average moisture content. Adequate cover of dense uncracked concrete to the reinforcement will minimize the chance of oxygen and moisture penetrating to the reinforcement. Provision of protection such as tanking on foundations will reduce contact with aggressive agencies. The choice of appropriate concrete mix proportions and well-shaped and graded aggregates will help to produce a dense durable concrete. These are all aspects of design that can play their part in influencing the useful lifespan of a structure.

Construction defects are of consequence where they permit external agencies, such as air and moisture, to enter the concrete and attack the matrix itself or the reinforcement. The most commonly occurring construction defects are failure to provide the design cover to the reinforcement, honeycombing due to low-workability concrete or inadequate compaction, and cold joints which may give problems in slip-form construction.

Durability of concrete is very much dependent on the proportioning of the concrete mix. Overbatching of water or under-batching of cement are construction errors which can lead to problems at an early age, particularly in aggressive environments, because they lead to a more permeable matrix. Sometimes water is added to the mix in countries with hot climates because of the rapid loss of workability at high temperatures. Adequacy of curing is also a significant contributory factor, as inadequate curing has a direct effect on the very important zone of cover concrete.

Early-age Defects

Concrete, at the time of placing in the mould, is a fluid mixture consisting mainly of water, cementitious material, sand and stone. The material remains fluid or plastic until the development of the cement hydrates which gradually give form and strength to the mass as described in Chapter 1. The solidifying process can take minutes or hours depending on the type of cement, the presence of admixtures and the ambient conditions. Moisture movements while concrete is in the plastic state can lead to cracks and other defects.

Plastic Shrinkage Cracks

The top surfaces of concrete pours are subject to evaporation and consequent loss of the mix water. The rate of evaporation is dependent upon ambient conditions such as temperature, wind speed and relative humidity. A method of assessing evaporation rates under a variety of conditions is given in ACI 305.[2.3] The water lost by evaporation is

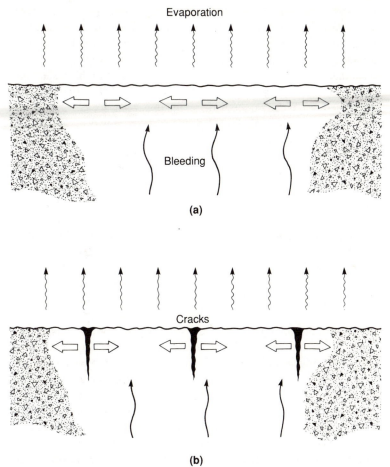

Fig. 2.1 Formation of plastic shrinkage cracks

Evaporation

Bleeding

(a)

Cracks

(b)

usually replaced by water rising to the surface of the concrete by the action of bleeding. Where the rate of evaporation from the surface exceeds the rate at which it can be replaced by bleeding, there is a local reduction in volume.

The movements (strains) associated with the reduction in volume are resisted by the concrete which lies immediately below and which is not subject to volume change. The restraint from the lower concrete causes tensile stresses to build up in the surface layer and, because the material still has very low strength at this early age, cracking can result. ACI 305 indicates that precautions against surface drying out and cracking should be taken if evaporation is likely to exceed $1 \, \text{kg} \, \text{m}^{-2} \, \text{h}^{-1}$. The process leading to plastic shrinkage cracking is shown diagrammatically in Fig. 2.1.

Although it would be expected that plastic shrinkage cracks would form in a random pattern, experience and research[2.4] have shown that this is not always the case. Cracks often form at right-angles to the direction of finishing or at 45° to the axis of a bay.[2.5] The crack pattern may be influenced by and follow the pattern of the top reinforcement. This is most likely to occur if the top reinforcement is close to the surface and can act as a stress raiser in the affected zone. Cracks caused by plastic shrinkage are typically quite wide on the upper surface (2−3 mm), but their width decreases rapidly below the surface. However, it has been reported that the cracks may pass through the full depth of the member[2.6] and the crack pattern is consequently repeated on the undersurface. The process leading to the formation of plastic shrinkage cracks does not explain, by itself, the production of full depth cracks, and it is probable that some subsequent occurrence, such as drying shrinkage, causes the plastic cracks to propagate. Plastic shrinkage occurs at a time when the matrix of cement hydrates is very weak. The cracks, therefore, tend to pass through the matrix rather than through the aggregate.

Plastic Settlement Cracks

The upward bleeding of water described above may be accompanied by a downward movement of the solid and heavier ingredients. This downward movement may be resisted by the top layer of reinforcement or by the formwork. In the former case, the layer of concrete above the reinforcement tends to become draped over the bars. If this occurs in the plastic rather than the fluid state the concrete may crack. In addition, the concrete may separate from the lower surfaces of the bars creating a void as shown in Fig. 2.2. When cracks are formed in this way, their pattern on the surface tends to mirror that of the reinforcement. The surface profile tends to be undulating with high points over the bars. In some cases this can be clearly observed when a straight edge is placed on the surface.

   In other circumstances, the downward movement of concrete can be restrained by the shape of the formwork. Plastic shrinkage cracks may occur at changes in sections, such as column heads or at splays at the base of contiguously poured walls, as illustrated in Fig. 2.3. Plastic settlement cracks on the upper surface of slabs or upstands tend to be of the order of 2−3 mm wide and taper to the level of the reinforcement. The cracks, as for plastic shrinkage, propagate predominantly through the matrix rather than through the aggregate.

Early Thermal Movement

As the chemical reaction between water and cement takes place, the concrete mixture stiffens and changes from the plastic to the solid state.

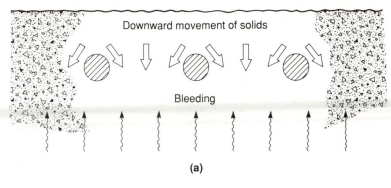

Downward movement of solids

Bleeding

(a)

Fig. 2.2 Formation of plastic settlement cracks

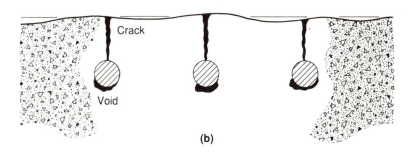

Crack

Void

(b)

The chemical reaction continues over an extended period but occurs at the maximum rate during the first few days. During this time, heat energy is released by the reaction at such a rate that it cannot all be dissipated rapidly through the surfaces of the concrete member. The temperature of the concrete rises. This effect is particularly pronounced in members of large cross-section or where insulating formwork is left in position. The rise in temperature may continue for several days, but usually the peak temperature occurs within 24 or 48 hours after placing.[2.7] Afterwards the concrete cools over a period of a few days until it reaches ambient temperature.

As for most other materials, the rise and fall in temperature are accompanied by expansion and contraction. The contraction does not pose any threat to the concrete unless some form of restraint is present. Unfortunately, in many practical situations, there is sufficient restraint, either external or internal, to lead to tensile stresses of such magnitude that cracking results. External restraint may occur because the concrete has been cast against a previously constructed base as in the classical case of retaining walls.[2.8, 2.9] Internal restraint occurs when the outer zone of concrete cools more quickly than the core. The reduction in volume of the surface layers is restrained by the inner concrete and cracks may develop. The cracks will later tend to close to a certain extent as

Fig. 2.3 Formation of plastic settlement cracks by formwork restraint

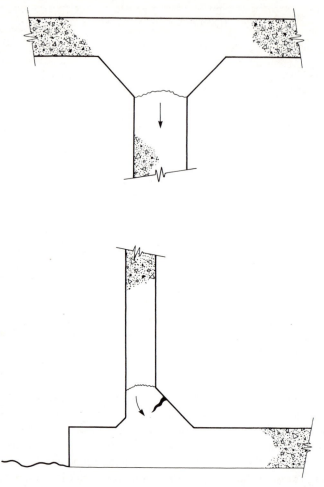

the member approaches uniform temperature. Temperature gradients can be particularly pronounced when shutters are removed during cold weather.

Shrinkage

The volume changes experienced by a concrete member do not end when it has returned to ambient temperature. Water is present in concrete mixes in much greater quantities than that required to take part in the hydration reaction with cement. This additional water is required during the construction stage to give the mix sufficient workability to enable it to be compacted easily into the shutters. The additional water is trapped in the mix as the concrete sets. Under normal conditions the moisture in the concrete is gradually released to the atmosphere.

The loss of moisture is accompanied by the reduction in volume known as shrinkage. As in the case of the reduction in volume associated with cooling from the hydration peak temperature, any restraint of the movement may result in tensile stresses which are sufficiently great to cause cracking. Shrinkage may also exacerbate existing cracks due to plastic shrinkage or plastic settlement. Extremely high shrinkage movements and excessive cracking have been reported for concretes made using unstable aggregates.[2.10]

Longer Term Problems

Concrete in service is exposed to a wide variety of environments and, because of its physical and chemical nature, may deteriorate as a result of this exposure. The structure of the matrix of cement hydrates contains pores and capillaries, and these provide passageways by which moisture and the atmosphere can penetrate the concrete. Cracking from any of the causes noted above assists this process.

Concrete is alkaline and is susceptible to reaction with acidic gases in the atmosphere or with acidic groundwater. Certain hydration products may take part in disruptive expansive reactions with sulphates in the aggregates or with those from an external source. In general, the higher the quality of the concrete, the less susceptible it should be to most forms of deterioration.

Freeze/Thaw

Concrete in exterior, damp situations is frequently in a condition where the pores in the outer zone are saturated with water. If it is exposed to freezing in this condition, ice crystals develop in the pores. As the crystals grow, they exert pressure on the surrounding concrete and may eventually cause small cracks to develop. The result is that the concrete breaks away from the surface in small flakes leaving the aggregate exposed. Under repeated cycles of freezing and thawing, the frost-shattering progresses deeper into the concrete and eventually aggregate pieces may become dislodged.

Freeze/thaw problems are likely to be encountered in situations where there are expanses of exposed horizontal concrete such as highways, flat roofs and runways. Attack may be patchy, being more intense at locations where water tends to run across the surface or at low points where water tends to accumulate.

Salt Weathering

The mechanism of salt weathering is very similar to that of freeze/thaw. In this case, salt crystals (commonly chlorides or sulphates) develop in the pores close to the surface. Crystal growth exerts pressure in a similar

Fig. 2.4 Damage
associated with
capillary rise
through structural
elements

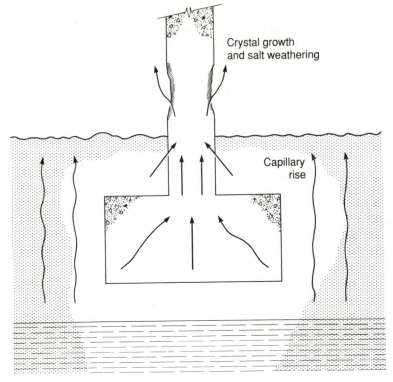

Crystal growth
and salt weathering

Capillary
rise

Saline groundwater

Fig. 2.4 Damage associated with capillary rise through structural elements

way to ice and again causes flakes of cement paste and fine aggregate
to become detached from the surface. This form of attack is common
in desert areas, and may also occur in marine structures. In the case
of deserts, salt from groundwater or damp soil is transported in solution
through vertical concrete members by capillary action. Above ground
level the moisture is drawn to the surface of the member. The moisture
evaporates and crystals of salt are deposited in the near-surface pores.
The result is a horizontal band of salt weathering which occurs just above
ground level (see Fig. 2.4).

In some cases the aggregate may be more susceptible to weathering
than the cement paste. Small discs of paste between aggregate pieces
and the outside surface are blown off due to the expansive action within
the aggregate. The features formed are called pop-outs. Eventually the
fragment of aggregate may disintegrate completely leaving a void (see
Fig. 2.5).

Sulphate Attack

Sulphates occur naturally in soils, rocks and groundwater in many areas
of the world. Gypsum (calcium sulphate) is present in some of the clay

soils in the south of England and also occurs in desert soils at locations where the water table is close to the surface. The more soluble sulphate salts, sodium sulphate (Glauber's salts) and magnesium sulphate (Epsom salts) may be present in some formations in the United Kingdom, but are more extensive in the alkali soils of North America.

Fig. 2.5 Pop-outs associated with expansive aggregates

Concrete can be attacked by sulphates in the soil and groundwater, or if they are present in the aggregate. If gypsum is present in the aggregate it may react with some of the products of reaction between water and cement (the calcium aluminate hydrates).

Calcium sulphate + tricalcium aluminate → tricalcium sulphoaluminate (ettringite) + calcium hydroxide

The ettringite occupies a greater volume within the concrete than the calcium aluminate hydrates. The expansion generates tensile stresses in the cement paste and as a result cracks develop in the concrete.

Sodium sulphate in groundwater reacts with the minerals present in concrete in two stages. The first stage is a reaction with calcium hydroxide to produce calcium sulphate and sodium hydroxide. The calcium sulphate then reacts with tricalcium aluminate as described above. If the sodium sulphate is replenished, as is the case where there is a flow of groundwater, the reaction can continue with further expansion.

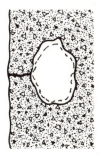

The reaction with magnesium sulphate in groundwater is also extremely disruptive. Magnesium sulphate reacts with both the tricalcium aluminate and calcium hydroxide to give tricalcium sulphoaluminate, calcium sulphate and magnesium hydroxide. This reaction lowers the pH of the pore water solution causing the calcium silicates in the cement paste to break down, releasing more calcium hydroxide. This calcium hydroxide reacts with further magnesium sulphate again lowering the pH, and so the reaction continues. If sufficient magnesium sulphate is available, the reaction continues until the calcium silicate structure in the cement paste is broken down completely and becomes soft and spongy. In severe cases the cement paste may be removed completely leaving only the aggregate.

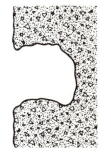

Sea water contains sulphates at concentrations which would be expected to cause damage to concrete. However, little damage has been attributed to the action of sea water on the concrete in structures. This is in sharp contrast to the severe effects reported due to the action of chlorides in sea water on reinforcement. It is thought that other ions in sea water have the effect of reducing the expansion that usually occurs when sulphates attack concrete. This may be because the products of sulphate attack are more soluble in sea water because of the presence of chlorides. An alternative explanation is that the magnesium hydroxide produced in the reaction blocks the pores of the concrete and hence reduces further penetration of the sea water.

In order to combat the effects of sulphate attack, concretes in sulphate-bearing soils are usually constructed using sulphate-resisting Portland cement. Sulphate-resisting cement has a low tricalcium aluminate content and hence less potential for expansive reaction.

Pyrites

Iron pyrites may be present in small quantities in some concrete aggregates. Fragments of pyrites near the surface may be oxidized leading to rust staining. The mineral iron pyrites consists of ferrous sulphide, FeS_2. In the presence of oxygen and moisture, ferrous sulphate and sulphuric acid are formed. The ferrous sulphate may be further oxidized to ferric hydroxide which is the cause of the orange-brown rust staining.

More serious problems have been encountered in south-west England with concrete and concrete blocks made using mine tailings containing pyrites and shaley impurities. This has become known as the mundic problem; 'mundic' apparently being the Cornish name for iron pyrites. The exact mechanism is not known, but appears to be related to the oxidation of pyrites and swelling of the shales. The result is a severe loss of strength in the concrete and it is understood that some structures may have been demolished as a consequence.[2.11]

Natural Acids

Some organic acids occur naturally in groundwater as a result of the decay of vegetable matter or the solution of atmospheric carbon dioxide. Lactic acid is present in sour milk and has sometimes caused damage to concrete in dairies. Acetic, butyric and lactic acids are produced by the fermentation reactions which take place in farm silage and these have caused damage to concrete and metal components of storage structures such as silos, clamps and pits. Sulphuric acid is sometimes produced by sewage in anaerobic conditions, or by the oxidation of pyrites. It may also be deposited in chimney flues.

If these acids come into contact with concrete they can react with the alkalis present in the cement hydrates to produce soluble products that can be removed by solution. In flowing water, the reaction products are carried away exposing fresh surfaces to attack. Under static conditions, the water adjacent to the structure may become saturated. If this happens the reaction ceases after only surface attack.

In the early stages, attack by acids generally results in the cement-rich paste being eaten away leaving the aggregate standing proud of the surface. The surface of the remaining paste may be softened. As the attack progresses the matrix may be removed completely leaving only the coarse aggregate. Where the coarse aggregate consists of limestone, this may also be attacked.

Desalinated Water

Water produced by desalination plants is extremely pure and contains only a low concentration of dissolved solids. Compounds are added to desalinated water to make it suitable for industrial and domestic use. If pure desalinated water is stored in concrete structures or conveyed in concrete pipes, it is capable of leaching out calcium hydroxide from the cement paste with an effect similar to the early stages of acid attack, described above.

Molluscs and Other Marine Borers

Some molluscs are capable of boring into hard-packed clay, coral, rocks and concrete. These molluscs form cylindrical holes slightly bigger than themselves and live in them for protection. Some species burrow by using their shells as augers. Sea urchins use a complex part of their body called Aristotle's lantern which is used in the function of both chewing and breathing. Marine borers have caused damage to a wide range of structures. A shellfish, *Martesia striata*, of the family commonly known as piddocks, has been recorded as boring through the 3 mm thick lead covering to an underwater power cable off the Florida coast. Sea urchins have been reported to cause damage to piles and underwater girders.[2.12] A shellfish of the mussel family, *Lithophago pirhuana*, has caused damage to concrete marine armour units in the Arabian Gulf.

Alkali Reactivity

Certain aggregates which contain silica are capable of taking part in reactions with alkali metal hydroxides in concrete. Alkali metals, sodium and potassium, are present in cement in small quantities. It is conventional to express the results of chemical analysis of cement in terms of the oxides, sodium oxide and potassium oxide. In order for simple comparisons between different cements to be made, the alkali content is usually expressed as total alkali content or equivalent sodium oxide content. To obtain the equivalent sodium oxide content, the potassium oxide content is factored by the ratio of the molecular weights of sodium oxide and potassium oxide, and added to the sodium oxide content.

Total alkali (as equivalent Na_2O)
$= Na_2O$ content $+ 0.658 \times (K_2O$ content$)$

The alkali content of cement depends on the materials from which it is manufactured and also to a certain extent on the details of the manufacturing process, but it is usually in the range of 0.4−1.6 per cent.

In concrete mixes there may be a contribution to alkali levels from other cementitious materials, such as pulverized fuel ash or ground granuated blast-furnace slag, which is present. Alkali metal levels may also be increased by contributions from sea water if, for instance, unwashed marine aggregates are used. They may also be increased in concrete while in service if there is exposure to a marine environment or to de-icing salts.

Alkali−Silica Reaction

During the hydration reaction the alkali metal compounds in cement react to produce sodium and potassium hydroxide which remain dissolved in the pore water in the hardened concrete. The alkali hydroxides can react with certain forms of silica present in some aggregates. The reaction produces various hydrous potassium and sodium silicates in the form of a gel. This gel absorbs water and, in the process, increases in volume and thus exerts pressure on the surrounding concrete.[2.13] The swelling pressure may be sufficient to cause microcracking which can eventually extend throughout the volume of the affected concrete. The gel may ooze out from cracks in the surface of the concrete. When the gel comes into contact with the atmosphere, it may carbonate and take on a white appearance.

Not all aggregates containing silica undergo reactions with alkalis, and not all reactions between aggregates and alkalis are disruptively expansive. It is usually the minerals which contain non-crystalline or cryptocrystalline forms of silica which are found to be reactive. These forms have large relative surface areas or a degree of disorder within their crystal lattice. Quartz, which occurs commonly in concreting aggregates, has a distinct and stable crystal form, and is stable in concrete under normal conditions. Early cases of expansive reaction in the United States were attributed[2.14] to the minerals chalcedony, crystobalite, opal and tridymite, which were present in the cherts, shales and siliceous limestones which made up the sand from which the concrete had been produced.

It is not necessary for the reactive minerals to be present in large amounts for extensive damage to be caused. In some cases, only a few per cent of highly reactive material in concrete has been found to have caused considerable damage to structures. Another factor is the size of individual reactive aggregate pieces. The cases of damaging reactivity reported in the United Kingdom have generally involved aggregate in the 1−5 mm size range.

Often the first external signs of alkali reactivity in a structure are short fine cracks on the surface radiating from a point. They occur adjacent to fragments of reacting aggregate and are caused by the outward swelling pressure (see Fig. 2.6). As time passes the cracks propagate and eventually join up to form map-like patterns. Typically, the crack pattern

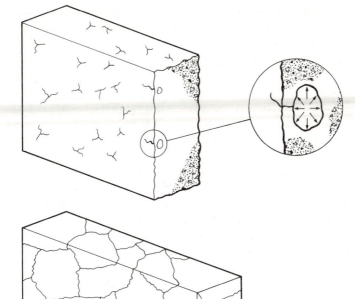

Fig. 2.6 Development of map cracking patterns associated with alkali reactivity

may appear within 3–5 years of construction. The macrocracks which appear on the surface are generally found to be of depth 25–50 mm and to occur roughly at right angles to the surface.[2.15] This occurrence of surface macrocracks is consistent with tension being generated in the outer layers by greater expansion within the core.

The crack patterns may be substantially modified by the presence of reinforcement prestressing or external restraints. The macrocracks may permit penetration by moisture and oxygen, resulting in corrosion which may also substantially modify the crack patterns. In some cases, the swelling may cause total disruption of fragments of aggregate close to the surface leaving pop-outs similar to those resulting from salt weathering.

In order for a disruptive reaction to occur three basic conditions have to be met.

1. A reactive aggregate must be present in sufficient concentration to sustain the reaction.

2. The pore solution in the concrete must contain alkali metal and hydroxyl ions in sufficient quantities to sustain the reaction.
3. Sufficient water must be present from an external source to be taken up by the gel and to cause it to expand.

It follows from Conditions (1) and (2) that the reaction, and hence the disruptive expansion, may cease when one of the reactive ingredients has been consumed or depleted. The state of the structure, in terms of extent of cracking due to alkali reactivity, may stabilize after several years.

From Condition (3) it follows that damage to concrete and the presence of visible cracks may vary considerably over a single structure. Those sections of a structure exposed to the prevailing rain-bearing wind, condensation or ponding will tend to be most severely cracked.

The tendency to crack in a particular case will depend on the rate and timing of the reaction, which is again dependent to a certain extent on the relative availability of the reactants. If, for instance, there is only a low concentration of reactive silica in the fresh concrete, the rate of production of gel may be slow. In this case the gel is able to dissipate through pores and microcracks, and insufficient pressure is built up to cause cracking. If, on the other hand, there is a very high concentration of reactive silica the reaction may be substantially completed before the concrete has set. The rate of gel production and growth after the concrete has hardened will be insufficient to produce cracking.

Alkali—Silicate and Alkali—Carbonate Reactions

An expansive reaction between alkalis in the pore water of concrete and certain aggregate containing silicates has been reported from Canada. The rock types involved are called phyllites and they have a layered structure like slate. A platy mineral related to vermiculite occupies the space between the main layers, which are of silicate minerals. A reaction between the alkalis and the vermiculite-like mineral dissolves away the interlayer material causing the rock to exfoliate. The silicate minerals are then able to take up water and expand. The expansion induces stresses in the concrete and hence cracking.

A reaction sometimes occurs between alkalis in concrete and some dolomitic limestones used as aggregates. Dolomitic limestones contain a proportion of magnesium carbonate in addition to the calcium carbonate found in common limestone. The magnesium carbonate in the rock is converted to calcium carbonate, magnesium oxide and alkali carbonates by reaction with the alkali metals. It is thought that this reaction does not cause expansion in itself, but that it exposes particles of desiccated clay which were within the dolomite crystals. When the clay minerals come into contact with water they expand and cause cracking of the concrete.

The alkali−carbonate reaction is relatively rare and has only been found in North America in some concretes containing dolomitic aggregates. The vast majority of sources of dolomitic limestone aggregates have not been found to be deleteriously reactive, and have been used successfully for the production of concrete.

Carbonation

The pore solution within freshly cast concrete is highly alkaline with a pH in excess of 12.5. The alkalinity is derived from calcium hydroxide and other compounds which are products of the hydration reaction of cement. Carbon dioxide and other gases in the atmosphere can penetrate concrete through its system of surface pores and capillaries. If water is present, carbon dioxide and other acidic gases can react with calcium hydroxide in concrete to form neutral compounds, such as calcium carbonate. The net effect is to reduce the alkalinity as follows:

$$Ca(OH)_2 + CO_2 \rightarrow CaCO_3 + H_2O$$

The process is known as carbonation. The outer zone of concrete is affected first, but with the passage of time carbonation proceeds deeper into the mass as carbon dioxide diffuses inwards from the surface. It is common to think of a carbonation 'front' progressing inwards from the surface, the front being the dividing line between carbonated and uncarbonated concrete. This may have come about because the common test for carbonation is to spray a freshly fractured surface with an indicator which changes colour at a certain pH. The indicator will give a distinct delineation at the particular pH value. The concrete below that pH is considered to be carbonated whilst that at a higher pH is considered to be uncarbonated.

This simplified concept may not be strictly true. If the indicator changes colour at pH 9 there will be a region ahead of the apparent carbonation front where carbonation has had some influence (see Fig. 2.7). None the less, the carbonation front is a useful concept and the depth of penetration is important for reasons discussed later.

The depth of penetration of the carbonation front is thought to be proportional to the square root of the time of exposure, if all influencing factors remain constant. When the time of exposure increases by a factor of four, the depth of penetration of the carbonation front doubles. For instance, if the carbonation front has reached a depth of 15 mm after a period of exposure of 10 years it will be expected to have reached a depth of 30 mm after a total exposure period of 40 years.

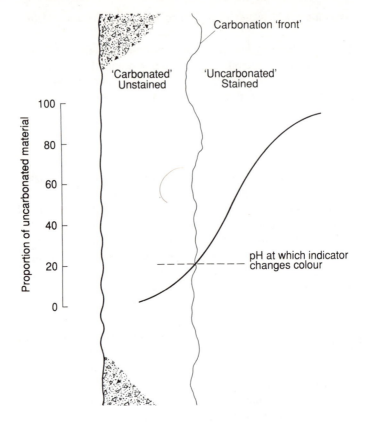

Fig. 2.7 Variation
of proportion of
carbonated
material near a
concrete surface

The rate of carbonation in individual cases is dependent on a number of factors. It depends on the 'quality' of the concrete, because carbon dioxide will have greater difficulty in diffusing through a better quality, less porous concrete with a high cement content and a low water/cement ratio. The carbonation depth will also tend to be greater at the location of cracks and other defects if they form a pathway by which the atmosphere can gain access (see Fig. 2.8).

The rate also depends on ambient relative humidity and the degree of saturation of the concrete. If the pores of the concrete are completely blocked by water, the carbon dioxide has difficulty in penetrating. On the other hand, carbonation could not take place in completely dry concrete because the reaction requires the presence of moisture to proceed. The rate of carbonation is influenced by the concentration of carbon dioxide to which the concrete is exposed. Carbon dioxide is present in the atmosphere at a concentration of approximately 300 parts per million but the concentration inside buildings may be higher.[2.16]

Taking the above factors into account, it can be seen that carbonation depths will not be uniform throughout a single structure. The depth will

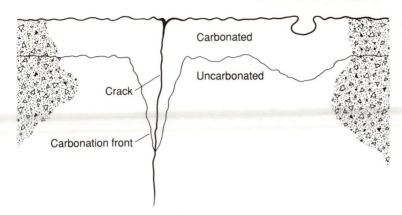

Fig. 2.8 Increase in carbonation depth at cracks and blowholes

Carbonated

Uncarbonated

Crack

Carbonation front

vary between indoor and outdoor locations, and there may be low carbonation depths on facades exposed to wet prevailing winds where the concrete remains in a saturated condition for longer periods.

Carbonation of concrete in itself is not a deleterious or disruptive process. It results in a slight reduction in volume, which may cause crazing or cracking of the outer layer as it is restrained by the unaffected inner concrete. It also changes the physical characteristics of the concrete and, therefore, may affect the results of surface tests, for example, rebound hammer. However, carbonation does have a major effect on the durability of reinforced concrete members for the reasons explained in the next section.

Corrosion of Reinforcement

The highly alkaline conditions inside concrete provide a passive environment for the reinforcement. It is thought that a thin layer of oxide forms on the surface of the reinforcement.[2.17] This oxide layer is stable in the alkaline pore solution and protects the steel from further corrosion. The steel is unlikely to corrode whilst passivating conditions remain. Well-compacted concrete also provides a physical barrier, reducing the penetration of atmospheric oxygen and moisture which are necessary to initiate and sustain the corrosion reaction. The corrosion reaction is electrochemical and creates measurable electrical potentials.

For steel in the alkaline environment of concrete the corrosion condition depends on both pH and potential. The situation has been summarized graphically by Pourbaix[2.18] as shown in simplified form in Fig. 2.9. It can be seen from the figure that reinforcement is in a passive condition at the high pH likely to be found in freshly cast concrete, but could start to corrode if the pH were lowered. As noted in the previous section, carbon dioxide gradually penetrates through pores and cracks in concrete causing a reduction in alkalinity. The reduction in alkalinity destroys

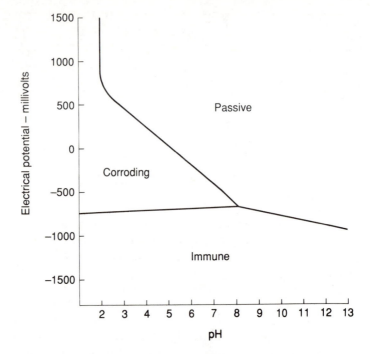

Fig. 2.9 Pourbaix diagram for steel in concrete

the passive environment and leaves the reinforcement in a condition where it is susceptible to corrosion. The passivation can also be destroyed by the presence of salts (commonly chlorides). Figure 2.10 illustrates the Pourbaix diagram for steel over a range of potentials and pH when chlorides are present. It can be seen that corrosion can take place even at the high pH within fresh concrete if chlorides are present.

In the past calcium chloride was added to concrete, in some circumstances, at the time of mixing as it acted as an accelerator. It was used to permit concreting to continue during cold conditions in winter. Calcium chloride was also commonly used in precasting works where its use resulted in an earlier formwork stripping time and hence gave the facility for greater production. Chlorides may also be present in aggregates. Many instances of corrosion from this cause have been reported in Middle Eastern countries.

Chloride salts sometimes enter concrete in service. This happens in marine structures such as docks and harbours, and to other structures close to the sea because of wind-blown sea spray. Many desert areas have high saline water tables, and in these regions moisture may rise up through the footings and columns of structures and, after evaporation, may reach concentrations sufficient to cause corrosion.

Salt is commonly spread on highways as a de-icing agent. As the ice melts it dissolves the salt, and on bridges the run-off may find its way through deck joints and be discharged onto the columns and piers of the

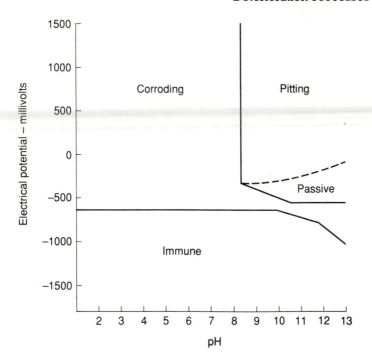

Fig. 2.10 Pourbaix diagram for steel when chlorides are present

substructure. Roadside structures, such as bridge abutments, are subject to spray from the wheels of vehicles, and it is known that the curtain of spray behind high vehicles may reach the soffits of bridges. The spray may well contain dissolved chlorides on occasions during the winter. Chloride from de-icing salts also finds its way into car-parking structures when it drips from the underside and wheel arches of cars.

When passivation has been reduced by carbonation or the presence of chlorides, corrosion of reinforcement may take place if sufficient oxygen and moisture are available. As stated previously, the corrosion process is essentially electrochemical in nature. It is driven by the occurrence of cathodic and anodic areas on the reinforcement surface. These may be caused by different local concentrations of salts or varying availability of oxygen along a bar's length. At the anode site, the iron dissociates to form ferrous ions and electrons:

$$Fe \rightarrow Fe^{++} + 2e^-$$

The electrons move through the metal towards the cathodic site but the ferrous ions are dissolved in the pore solution.

At the cathodic site, oxygen in the pore solution combines with the electrons to form hydroxyl ions:

$$2H_2O + O_2 + 4e^- \rightarrow 4OH^-$$

The ferrous and hydroxyl ions move in opposite directions through

Fig. 2.11 Corrosion
of reinforcement
and resultant
cracking

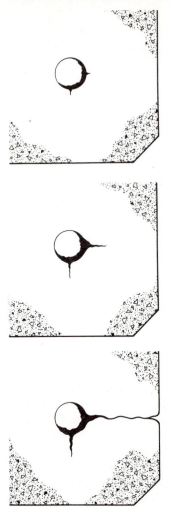

the pore solution, and when they meet ferrous hydroxide is precipitated:

$$Fe^{++} + 2OH^- \rightarrow Fe(OH)_2$$

Corrosion usually starts first on the outer faces of the bars as the concrete in contact with these areas is the first to become carbonated and these outer faces are closest to external sources of chlorides, moisture and oxygen. The precipitated corrosion products occupy a bigger volume than the original steel. As the corrosion products build up they exert a gradually increasing pressure on the concrete until it cracks. In most cases the crack propagates from the bar to an adjacent surface, and the cracks on the surface follow the line of the underlying reinforcement (see Fig. 2.11). As corrosion continues and cracks propagate, corners of beams and columns may become detached and spall.

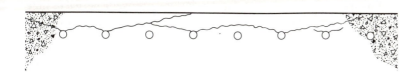

On slabs with particular ratios between bar spacing and cover, the cracks may form at a shallow angle to the surface and be intercepted by cracks from adjacent bars before reaching the surface. The result is that a plate of concrete above the reinforcement becomes detached from the main body with few surface cracks only intermittently visible on the surface, as shown in Fig. 2.12. This phenomenon is known as lamination or delamination. The concrete sounds hollow and dull when the surface is struck with a hammer.

Corrosion of reinforcement may also take place in chloride-affected concrete when little oxygen is available. In this case, the corrosion is localized and forms deep pits. The volume of corrosion products generated may be insufficient to cause surface cracking, and it is possible that there could be severe loss of cross-section of some bars with very little prior warning from visible signs on the surface. It is thought that this mechanism can occur in saturated concrete.

The corrosion process is electrochemical in nature and corrosion activity is accompanied by measurable electrical potential differences across the surface of a structure with some areas of reinforcement acting as cathodes and other areas acting as anodes.

High Alumina Cement Concrete

High alumina cement (HAC) was first developed in continental Europe and was used originally mainly in structures exposed to sulphate-bearing soils and groundwater. Concrete made with HAC develops high early strengths, and this property made it attractive to the operators of precast works even though it was more expensive than ordinary Portland cement. The higher cost was more than offset by the ability to strip the formwork on the day after casting, thus permitting rapid re-use and increased rates of production.

The use of HAC in this way was common in Britain in the 1950s and 1960s when many precast prestressed beams and floor slabs were manufactured using HAC. Concrete made from HAC tends to be somewhat darker than concrete made from ordinary Portland cement. This is a useful aid to recognition along with a knowledge of the typical circumstances under which it was used and the types of members in which HAC was employed.

The principal product of hydration of HAC is monocalcium aluminate decahydrate. This compound is not stable at normal temperatures found in buildings and structures, and it slowly converts over a period of time to the more stable mineral forms, tricalcium aluminate hexahydrate and gibbsite. The conversion is generally accompanied by a significant loss of strength, and the resulting concrete may be more porous and more susceptible to chemical attack.

The loss of strength of high alumina cement concrete members was a contributory factor in the collapse of two roofs in Britain in 1973. At that time, the problem was investigated extensively and HAC was subsequently written out of British standards for the structural use of concrete. However, a large number of structures were built when HAC was in popular use, and members containing HAC may be encountered when buildings are repaired or refurbished.

Fire Damage

Concrete construction is relatively good at resisting the effects of damage by fire. Concrete is a poor conductor of heat and, in fact, the thermal conductivity is reduced as the temperature increases. High temperatures within members may be limited to the outer layer if exposure to fire is not prolonged.

If concrete is exposed to a rise in temperature, the water contained in its pores and capillaries is first driven off. As the temperature increases above 100 °C some of the water, combined with the calcium silicate hydrates in the cement paste, is also lost and this desiccation is accompanied by a reduction in strength. The degree of desiccation and strength reduction are dependent on the temperature reached and the length of exposure.

At temperatures between 300 and 400 °C there is a more pronounced chemical change in the cement paste. The calcium silicate is converted to calcium oxide and silica, and the concrete experiences a further appreciable loss of strength. In many concretes exposed to temperatures in this range, there is also a distinct colour change. The concrete takes on a noticeable pinkish tinge.

The damage caused to concrete members by fire is generally one of two types. Depending on the moisture content, explosive spalling may occur in the early stages of exposure. There is a series of violent disruptions each of which removes a shallow layer of concrete from the surface over a localized area. The other type of damage commonly experienced in concrete members subjected to fire is the rounding of arrises on columns and beams. Internal cracks develop almost parallel to the faces and eventually a plate of concrete becomes detached. The detachment often occurs at the position of the outside face of the reinforcement. Considerable damage may also occur because of the thermal shock experienced by the concrete as water is sprayed onto the structure during fire-fighting. This may be in the form of cracks or further spalling.

The steel in reinforcement retains its strength to much higher temperatures than concrete. At 400 °C, most reinforcement retains at least 90 per cent of its strength. Thereafter, there is a reduction in strength until at 700 °C only about 20 per cent of the original strength remains. However, the steel regains strength on cooling typically to at least 70 per cent of original strength after cooling, from 700 °C.

The effect of high temperature on prestressing steel can be much more significant. A 50 per cent reduction in strength can be experienced at 400 °C. However, the effects on the load carrying capacity of the member may be greater than this loss would suggest because of simultaneous reduction in the elastic modulus of the concrete, relaxation of the steel due to creep and the thermal extension of the strand or wire.

References

2.1   Paterson A C 1984 The structural engineer in context. *The Structural Engineer* **62A (11):** 335−42

2.2   Rasheeduzzafar, Al-Gahtani A S, Al-Saadoun S S 1989 Influence of construction practices on concrete durability. *American Concrete Institute Materials Journal* **Nov-Dec:** 566−75

2.3   ACI Committee 305 1982 *Recommended Practice for Hot Weather Concreting* ACI Standard 305-82, American Concrete Institute, Detroit

2.4   Shaeles C A, Hover K C 1988 Influence of mix proportions and construction operations on plastic shrinkage cracking in thin slabs. *American Concrete Institute Materials Journal* **85 (3):** 495

2.5   Concrete Society 1982 *Non-structural Cracks in Concrete* Technical Report No 22, The Concrete Society, London

2.6   Lerch W 1958 Plastic shrinkage. *Journal of the American Concrete Institute, Proceedings* **54 (3):** 797−801

2.7   ACI Committee 207 1990 Effect of restraint, volume change and reinforcement on cracking of mass concrete. *American Concrete Institute Materials Journal* **May-June:** 271

2.8   Hughes B P, Miller M M 1970 Thermal cracking and movement in reinforced concrete walls. *The Institution of Civil Engineers, Proceedings* Paper No 72545, pp 65−86

2.9   Hughes B P 1973 Early thermal movement and cracking of concrete. *Concrete* **May** London: 43−6

2.10   Building Research Establishment 1968 *Shrinkage of Natural Aggregates in Concrete*. Digest No 35, The Building Research Establishment, Garston, United Kingdom

2.11   Anon 1990 Suspect aggregates still going into concrete. *New Builder* **5 July** London, p. 6

2.12   Gardiner M S 1972 *Biology of Invertebrates* McGraw-Hill, New York

2.13   Hobbs D W 1984 Expansion of concrete due to alkali−silica reaction. *The Structural Engineer* **62A (1):** 26−34

2.14   Stanton T E 1940 The expansion of concrete through reaction between cement and aggregate. *Proceedings of the American Society of Civil Engineers* **66:** 1780−811

2.15   Hobbs D W 1988 *Alkali−silica Reaction in Concrete* Thomas Telford, London

2.16   Richardson M G 1988 *Carbonation of Reinforced Concrete: Its Causes and Management* Citis Ltd, Dublin

2.17   Sagoe-Crentsil K K, Glasser F P 1989 Steel in concrete: Part I. A review of the electrochemical and thermodynamic aspects. *Magazine of Concrete Research* **41 (149):** 205−12

2.18   Pourbaix M, translated by Franklin J A 1966 *Atlas of Electrochemical Equilibria in Aqueous Solutions* Pergamon Press, New York

3 Planning an Investigation

Investigatory work is essential before undertaking renovation or repair of a concrete-framed or clad building or a civil engineering structure. In the case of renovation, or change of use, it is usually necessary to determine the details of the design and the current strength of the concrete and the state of the reinforcement, to assist in assessing the load carrying capacity. In the case of a structure where some form of deterioration has taken place, the causes must be established so that appropriate remedial and protective work can be designed and specified.

The most common reasons[3.1,3.2] for carrying out appraisals or assessments of structures are:

(a) Some form of deterioration is apparent. Examples of this are cracking, spalling, staining, loss of surface or erosion of the concrete.
(b) There is apparent structural weakness or distress. Evidence of this would be shown by large deflections, cracking or spalling.
(c) A change of use of the structure is being considered.
(d) The structure has been generally or locally damaged by an accidental loading such as impact, explosion, fire, earthquake or settlement.
(e) The structure is changing ownership.
(f) A defect is suspected in the structure.
(g) The appraisal forms part of a routine maintenance programme.

This chapter will discuss overall planning and other general aspects of investigations and the following chapter will describe investigatory techniques and methods in detail.

Safety

The safety of inspection personnel is of prime concern at all stages of a survey and needs to be kept constantly in mind during the planning process. Some aspects of safety during surveys are discussed in a booklet produced by the Health and Safety Executive in the United Kingdom.[3.3]

The booklet is not primarily intended for concrete structures but the principles are generally applicable to structures of all types.

Inexperienced inspectors should not be permitted to work in or around buildings on their own. In circumstances where work is carried out alone, a system of regular reporting to a responsible person should be instituted. This allows appropriate action to be taken in the event that a routine report is not received by the specified time. When structures involving industrial processes or machinery are being inspected, it is advisable for the inspector to be accompanied by a member of staff from the operating company. This person should be in a position to make the inspector aware of any potential hazards.

Consideration also needs to be given to safety equipment and clothing appropriate to the circumstances. In most cases it is a wise precaution to wear safety hats and boots. Overalls and waterproof outer clothing are necessary in dirty or wet conditions. In some circumstances dust masks may be required. There may be dust deposits, such as asbestos, in any structure as this material has been used in the past in insulation and cement-bound products such as corrugated sheeting. Dust deposits are particularly likely to occur in warehouses and chemical plants.

If scaffolding, cradles or similar access equipment are to be provided by others for the inspection, they should not be used until the necessary safety certificates have been checked. This applies also to permanent cradles on buildings. Fire escapes and fixed ladders should be carefully examined before being mounted. They may have been poorly maintained, paint may mask corrosion and the fixings may be in poor condition. It is advisable to start from the bottom and work upwards assessing each stage ahead before proceeding.

Before entering a building that has been unoccupied for many years, or has been damaged by fire or some other cause, a careful external reconnaissance should be carried out. The aim is to check the integrity of the structure as far as is possible by a visual assessment of the verticality of columns and walls, and the line of horizontal members. If this preliminary inspection indicates that the overall condition of the structure is stable, the structure can be entered with caution. Again, a staged approach is advisable.

Each floor in turn, including the ground floor which may be suspended over a basement, should be assessed for soundness before proceeding. Basements, sewers and other similar confined spaces represent a particular potential hazard because of the possibility of accumulation of poisonous or asphyxiating gases. Inspection personnel should be familiar with relevant safety procedures.[3.4]

It should be remembered that structures which have been unoccupied and open for some time may have become a refuge for wild animals, reptiles or insects. This is not usually too much of a problem in temperate climates, except that rats can be a cause of the spread of leptospiral jaundice (Weil's disease), and accumulations of droppings from any animal may constitute a health hazard. In tropical and sub-tropical climates, the possibility of encountering poisonous snakes, spiders and other insects should be kept in mind. Another aspect that needs to be kept in mind is that unoccupied structures may have been used as a meeting place by drug users, and hypodermic needles may have been left behind.

Overall Plan

Irrespective of the underlying reason for the appraisal or assessment, it usually involves some form of survey. The survey has the objective of providing sufficient information to allow a realistic assessment of the structure to be carried out. The assessment may be of the load carrying capacity of the structure, its condition in terms of the various types of deterioration described in Chapter 2, or both of these. The individual activities and techniques used in the survey will be different depending on the type of assessment required but, in most cases, the overall plan will be the same. There will be an initial phase when general dimensions and global properties are determined, and a later phase when particular locations are examined in greater detail. These principal stages in the investigation of a concrete structure are illustrated for both load assessment and condition assessment in Fig. 3.1.

Initial General Phase

As can be seen from Fig. 3.1, assessments of both load-carrying capacity and structure condition include a walk over survey and research of existing records in the initial general phase.

Walk Over Survey

The first objective in most investigations, no matter what their size or scope, is to gather readily available information on the structure in question. As part of this objective, in many cases the first activity is an initial walk round inspection. Sometimes drawings or other details may be available before the first visit to the site and these will provide a useful introduction, although some caution is necessary as these records may not accurately represent the construction as built. Otherwise the walk round inspection provides the opportunity to make initial assessments of the structural form and to take general note of the locations

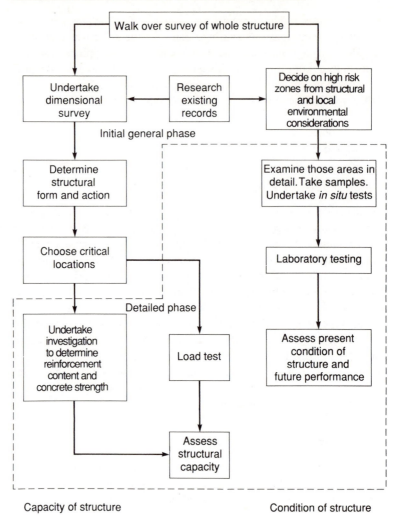

Fig. 3.1 Flow chart
indicating stages
in an ivestigation
of a reinforced
concrete structure
(after Fookes,
Pollock and Kay[3.5])

of any deterioration. It is also useful to note any problems likely to be
encountered with access to critical locations during the subsequent more
detailed survey, including the need to remove cladding or other finishes.
This first walk round inspection is very important as the impressions
gained will often set the tone and direction of the later stages of the
survey.

Existing Records

The second activity carried out early in the assessment process is the
attempt to find existing records of the structure. These records can prove

invaluable in reducing the amount of investigatory work that is necessary. If the owner or occupier of the structure either does not have drawings or not know of their whereabouts, it is worth checking with the local record office and the fire authority. Other possible sources of information on buildings are given in an appendix to a report by the Institution of Structural Engineers.[3.6] However, the recorded details should always be subject to checks at site. There may be discrepancies between what is shown on drawings and what was actually constructed. The drawings or specifications will show the design and specified strengths of the materials. Those actually achieved in the construction could be significantly different.

Detailed Phase

Once the preliminary walk-over survey has been completed and the search for existing records has been put in hand, the detailed phase can be planned with more confidence. The extent and scope of this work will depend on many factors, including the structural form, the quantity and quality of existing records, the degree of anticipated change of use and alteration of the structure, and the extent of deterioration. The individual activities will differ depending on whether an assessment of structural capacity or condition is being undertaken. However, problems of provision of access and rate of sampling are common to both types of assessment.

Access

The provision of access requires consideration early in the planning phase. The form of access should be such that it provides a suitable and safe platform for the planned activities and equipment, and permits the surveyors/inspectors to obtain a close view of critical areas of the structure. In many cases, ladders and lightweight scaffold towers will be found to be adequate. Many tall buildings have their own cradles for window cleaning or routine maintenance, and these can be used for inspection. Items of mobile equipment such as scissor lifts, satellite towers and cherry-pickers are widely available. The latter can be expensive, depending on size and reach, but may often provide the only realistic means of gaining access to difficult locations. Purpose-built equipment mounted on road vehicles or railway bogies has been developed to provide access to the superstructures of bridges. There are specialized companies who undertake surveys of façades of buildings using abseiling techniques. This method can provide an economic means of obtaining information.

Sampling

At the planning stage, the number of samples required has to be assessed. In many instances, it is best to approach this problem from a statistical point of view,[3.7] but this may not always be appropriate or possible.

The usual objective in taking and testing samples from a structure is to be able to arrive at estimates of the overall mean value of some property such as strength. In the planning stages of an investigation it is necessary to determine how many samples are required to give a reasonably precise estimate of the mean value. The procedure below uses strength as an example of the property to be determined, but is valid for other properties such as cover to reinforcement, carbonation depth and chloride content, so long as the property varies in a random manner and an unbiased sample can be obtained.

It is first necessary to break down the structure into notional units and to decide on the size of unit which each sample is to represent. This is fairly straightforward in the case of beams, columns or precast cladding units where one or two samples or tests could be taken to represent a member. However, it may need a little more thought in the case of large planar members like floor slabs and walls. In these cases, the unit size could be defined by reference to the volume of concrete contained, structural grid layout, layout of pours or an arbitrary grid. As an example, a factory ground floor slab of dimensions 60 m × 30 m might be sampled on the basis of a unit size of 2.5 m × 2.5 m. There would be 288 units of this size in the floor and their strength could be represented by the compressive strength of cores taken from their centre. If the full population of 288 units were tested the results should follow a normal distribution about a mean value. However, estimates of the mean can be obtained and, in most practical cases, are likely to be obtained from a much smaller number of randomly chosen samples.

The distribution of results from the full population and a randomly chosen sample are shown in Fig. 3.2. The average compressive strength for the full population of 288 units is $\bar{x}_p$ and the average compressive strength of the sample is $\bar{x}_s$. The difference between these two values is the sampling error of the mean compressive strength

$$\bar{x}_p = (\bar{x}_s \pm \text{sampling error}) \qquad [3.1]$$

Statistical theory indicates that the sampling error (SE) is given by the standard deviation of the whole population divided by the square root of the number of samples

$$SE = \sigma_p/\sqrt{n} \qquad [3.2]$$

In this case, the population standard deviation is not known but the

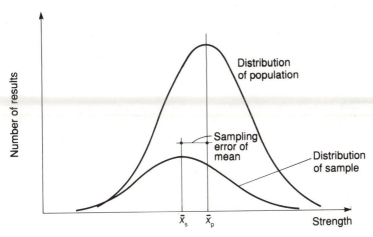

Fig. 3.2 Distributions of results for population and sample showing sampling error

sampling error can be determined from the sample standard deviation multiplied by a variable factor t, so that:

$$SE = t\sigma_s/\sqrt{n} \qquad [3.3]$$

The variable t is known as 'student's t' and can be obtained from standard tables on the basis of sample size and the confidence limits desired. The standard tables are usually related to the number of degrees of freedom, which is one less than the sample size.

Equation 3.3 can be rearranged in the form below, which gives the number of samples required to arrive at an estimate of the mean to the desired degree of accuracy:

$$\text{Number of samples, } n = (t\sigma_s/SE)^2 \qquad [3.4]$$

Returning to the example of the floor slab with 288 notional units mentioned above, if the mean strength is to be determined with an error of less than 2.5 N mm^{-2} and with a confidence limit of 99 per cent, the number of samples required can be calculated. A sample standard deviation has to be assumed. Assuming that the standard deviation is 3.0 N mm^{-2} gives:

$$n = (3.5 \times 3.0/2.5)^2 \qquad [3.5]$$

Standard tables show that for a 99 per cent confidence limit, t is less than 3.5 if the sample size is greater than 8. Evaluating eq. 3.5 leads to a first estimate of the sample size of 18.

The 18 samples would be chosen by assigning each of the 288 units of floor a unique number and using random number tables or a random number generating computer program. A core would be taken from the centre of each chosen bay and tested. There would have to be some pre-arranged procedure for choosing additional sample locations should one or more of the chosen bays be inaccessible for some reason. After testing the samples, the actual standard deviation of the results would be used to check that the desired accuracy had been achieved and further samples chosen at random and tested if necessary.

It is known that the concrete strength may vary quite widely within an individual member[3.8] because of the differences in compaction or curing achieved at different locations. For instance, the compressive strength of concrete at the base of a column may be greater than that at the top because of differences in the compaction due to the head of wet concrete. This factor needs to be kept in mind when choosing sample sites; for instance, by choosing the same relative position on each member.

Structural Capacity

There are several factors which make the investigation of reinforced concrete structures particularly challenging. Among these are:

1. A reinforced concrete member derives its strength from both the reinforcement and the concrete. The strength developed by a member is also dependent on the workmanship during construction and its maintenance history. Hence the load-carrying capacity of the re-inforced concrete member is not directly dependent on the external dimensions alone.
2. Concrete as a material has a wide strength range.
3. The strength properties of concrete are time-related.
4. Attack by external agencies or internal chemical processes may result in substantial strength reductions.

These factors have to be kept in mind when undertaking assessments of structural capacity. One particular consequence is that the strength of concrete in a structure may vary from location to location.

In order to carry out an assessment of the structural capacity of an existing reinforced concrete member it is necessary to determine:

(a) member dimensions;
(b) concrete strength; and
(c) reinforcement strength, content and location.

It is not usually possible or necessary to determine member dimensions and reinforcement size and position at all locations. In normal circumstances, they are determined only at a relatively small number of selected critical locations or sections, which are chosen using knowledge of structural action and other factors. As an example, in members subject to bending, such as beams, the sections would be chosen at positions of maximum bending moment such as mid-span and near the supports. Sections would also be chosen to try to determine the possible causes of any obvious distress such as cracking or excessive deflection.

Strength

A range of methods of assessing strength of concrete is available including taking and testing cores, and other less destructive techniques. The details of these techniques will be given in the next chapter. At the planning stage it is only necessary to consider the tests in broad outline so that the choice of an appropriate method can be made. Table 3.1 lists the available tests with some remarks on their applicability.

Reinforcement

As stated above, in order to estimate structural capacity it will be necessary to determine the type, amount and position of reinforcement. If the bars are not closely spaced and the arrangement of steel is not too complicated, the bar spacing and depth can often be determined using a cover meter, as described in the next chapter. It may also be possible to obtain an estimate of bar diameter using this instrument.

A certain amount of opening up is almost inevitable to check the results, and the locations for this will also have to be decided at the planning stage. The positions are decided mainly from a preliminary consideration of the structural form and judgement of the locations of the critical sections, where it is important that details are known with some certainty. In some structures breaking out may be inadvisable without propping, and this requirement also has to be taken into consideration at the planning stage. An alternative method of determining reinforcement content might be to use radiography or some other non-destructive technique if satisfactory information can be obtained by these methods. These techniques are also discussed in more detail in the next chapter.

Load Testing

Load testing is sometimes used as an alternative method of assessing structural capacity. Load tests are usually carried out for one of the following reasons:

1. There are still doubts about the satisfactory performance of the structure under load after a survey and local testing.

Table 3.1 Test methods for assessing strength of concrete in structures

Test method	Applicable standards	Remarks
Coring	BS 1881: Part 120 ASTM C 42	Gives direct assessment of strength and most accurate result. Samples can be obtained at depth. Partially destructive — core holes need to be repaired. Relatively expensive. Noisy.
Pull-out	BS 1881: Part 207 ASTM C 900	Need for calibration. Close to surface method. Partially destructive — test sites need to be repaired. Intermediate cost.
Pull-off	BS 1881: Part 207	Need for calibration. Close to surface method. Partially destructive — test sites need to be repaired. Intermediate cost.
Penetration	BS 1881: Part 207 ASTM C 803	Need for calibration. Close to surface method. Partially destructive — test sites need to be repaired. Intermediate cost. Nosiy.
Surface hardness	BS 1881: Part 202 ASTM C 805	Need for calibration. Surface method. Non-destructive. Test location must have a smooth surface. Relatively inexpensive. Relatively quiet. Quick.
Ultrasonic pulse	BS 1881: Part 203 ASTM C 597	Need for calibration. Gives average strength over path length. Non-destructive. Relatively inexpensive. Quiet. Relatively quick.

2. It is difficult or impossible to determine adequate information about the structure and its materials.
3. Verification of structural analysis in cases where the complexity of the structural form does not lend itself to rigorous analysis.
4. Deficiencies in detail, material or construction are suspected and such deficiencies would mean that the normal procedures or assumptions on which structural analysis is based were not appropriate. Examples of this would be where beam/column junctions had been incorrectly detailed or materials did not conform to specification in some way.

Clearly load tests should only be carried out if a reasonable interpretation of structural adequacy can be obtained from such a test and the influence of adjacent structural members can be taken into account when assessing the results.

In the context of assessment and renovation of structures, load tests are likely to be service load tests or overload tests. Service load tests may be carried out to check whether the structure meets deflection and crack-width criteria, or to assess some aspect of structural behaviour (e.g. interaction between beams and columns). Overload tests are carried out to check that an adequate margin of safety exists when the structure carries the load anticipated in service, i.e. the test load is the service load times a chosen factor of safety.

Load tests require careful planning because they are expensive and there is rarely the opportunity to repeat them. At the planning stage for a load test, the preliminary processes described earlier in this chapter should be followed. The structure should be visited for an initial assessment of its condition and other matters likely to have a bearing on the smooth running of the load test.

It will be useful to undertake approximate calculations before detailed planning gets under way. The calculations give a better understanding of the structural action and allow estimates of the loading and maximum deflections to be made. Some thought should also be given to the way in which the results will be interpreted and the tests planned so that they yield sufficient relevant information for the required calculations.

The planning for the load test should always be undertaken on the basis that the structure or the section under test may fail. Particular caution is necessary when dealing with structures where there has been obvious deterioration or where the strength of the concrete or reinforcement is likely to have been reduced; for example, high alumina cement concrete structures or structures affected by fire. It may be necessary to provide bracing to protect adjoining properties or to restrain lateral movements of free-standing structures. In the case of floor slabs it is usual to provide a catch scaffolding or loose props with sufficient clearance to accommodate the anticipated deflections.

In the event of a collapse, the load, including impact, is likely to exceed the capacity of the floor below and, therefore, it is usually necessary to prop through several floors. The design of the propping system needs to take into account that an additional scaffold system will probably also have to be provided to support instruments such as dial gauges. It will also be necessary to provide sight paths to allow the gauges to be monitored without the need for personnel to enter the dangerous region beneath the loaded structure.

The layout of the structure under test may make it difficult for the personnel supervising the loading operations to maintain contact with the staff reading the instruments. The provision of field telephones or two-way radios should be considered.

Loading

The most common forms of loading employed for load tests are water, bricks or blocks, sand and metal weights. Water is convenient to use and can be dispersed fairly quickly if the need arises. The water is usually contained in purpose-built tanks of rubber or plastic sheeting or in small individual vessels such as plastic barrels. The containment arrangement needs to be carefully considered as leakage can seriously damage finishes, electrical equipment and other items which remain in the building during the test. When tanks with large surface areas are used, it will be necessary to check the water depths at several positions as deflections of the structure may cause significant variation over the tank.

The use of dead weights such as bricks, blocks and metal weights can be expensive because placing them in position is time consuming and labour-intensive. Careful consideration needs to be given to the considerable logistical problems that can be involved in transporting the large numbers of bricks into their required position. This will have to be accomplished without causing unacceptable concentrations of load at any position.

The load provided by bricks and blocks is dependent on their moisture content. They should, therefore, be protected from the elements and a means of checking the weight of samples should be available on site. The load is provided by forming small discrete stacks. The stacks should be of stable proportions to avoid toppling as the floor deflects. They should also be sufficiently far apart to prevent arching.

Instrumentation

The most common forms of instrumentation employed during load tests are dial gauges and strain gauges. During the planning stages, it is necessary to assess the position and magnitude of the maximum readings so that the instruments can be appropriately positioned and instruments with the correct range can be obtained. Additional gauges should be placed near the supports of beams so that axial shortening of columns can be taken into account and the true member deflections can be determined. It is necessary to provide a rigid support system for the dial gauges which is independent from the catch scaffold and which will be unaffected by the work going on in the area. At some stage it will be necessary to correct for thermal movements of the support system. This can be done by either taking a series of dummy readings with no load over a daily cycle, or by suspending an Invar plumb bob from part of the structure under test.

Condition of Structure

The initial walk-round inspection may provide clues to underlying causes of deterioration and hence give a guide to the forms of testing which are appropriate. Various deterioration mechanisms and the surface features which result from them were discussed in Chapter 2. These surface features can be used during the initial inspection of a structure as a guide to its condition. To use a medical analogy, the visual signs are symptoms of a possible underlying disease. However, just as in medicine, symptoms are not always sufficient to make an accurate diagnosis and it is wise to undertake some tests to confirm preliminary findings. Nevertheless, the outward signs can be useful in that they give a general guide to the type of testing that may be required. Some frequently encountered defects are listed in Table 3.2 with their possible

Table 3.2 Some common defects and their possible causes with suggestions for confirmatory testing

Observed defect	Possible causes	*In situ* sampling or testing	Laboratory testing or other action
Cracks along line of reinforcement	Plastic settlement	Coring	Inspection
	Reinforcement corrosion	Coring	Inspection Carbonation depth Chloride content on incremental samples Cement content
		Dust sampling	Chloride content on incremental samples
		Carbonation depth Cover meter Hall-cell	Cement content
Cracks on top surfaces of slabs	Plastic shrinkage	Coring	Inspection
Map cracking Pop-outs	Alkali reactivity	Coring	Petrographic examination
Other cracks	Early thermal movement	Crack mapping may eliminate some of	Check history
	Plastic settlement and shrinkage	the possible causes	
	Inadequate capacity	Coring	Check design Strength tests Cement content
		Cover meter survey *In situ* strength determination	
Spalling	Reinforcement corrosion	Coring	Inspection Carbonation depth Chloride content Cement content
		Dust sampling Carbonation depth Cover meter Half-cell	Chloride content on incremental samples Cement content
	Fire damage	Coring	Inspection Petrographic examination
Loss of surface	Salt weathering Frost damage		Coring Chloride content Sulphate content
Cracking and disintegration of foundations	Sulphate attack	Lump samples	Sulphate content

causes and suggestions for confirmatory tests. The timing at which defects first occur can also be a useful indication of their origin[3.9] and this should be taken into account when accurate information is available.

At the planning stage, it will be necessary to choose the locations where the various tests are to be carried out and samples are to be taken, but these may have to be modified as the investigation proceeds. Test and sample locations will usually be chosen because they are in the vicinity of some of the defects mentioned above or because they are typical of many areas on the structure. Some testing should be carried out, if possible, on areas which are free from obvious signs of deterioration for comparative purposes.

Recording Results

It is essential to have some convenient method of recording the locations of sampling and testing when the site work is undertaken. In order to achieve this it is necessary to set up a simple grid or other referencing system for the whole structure. In the simplest cases, this need only be a numbering system which includes each element. Plans and elevations of areas of the structure which are to be studied in detail can be extremely useful and may be used as field sheets for recording the results of *in situ* tests such as cover meter or half-cell surveys as illustrated in Fig. 3.3. A wide range of electronic data-loggers is also available. These can be interfaced with many of the instruments commonly used in investigations, and they provide a convenient means of recording and processing the results of large surveys.

At the planning stage it is also worth setting out standardized descriptions or classification systems for degrees of deterioration. The form and degree of detail will depend on the scope of the survey and its underlying intention. ACI Committee 201 has given a helpful checklist for use when undertaking a survey of concrete structures.[3.10] The list contains many items which relate back to the time of initial construction and is intended partly to be used in progressively recording the history of a structure. However, it also forms a useful guide to the kind of information that should be collected as part of a survey or investigation at any stage during a structure's life. The Committee's report also has an appendix in which many descriptive terms associated with the durability or deterioration of concrete are defined.

When undertaking a fairly basic visual survey of a large number of structures at different locations the following global classification system has been found to be useful:

Category A
No visible defects or previous repairs.

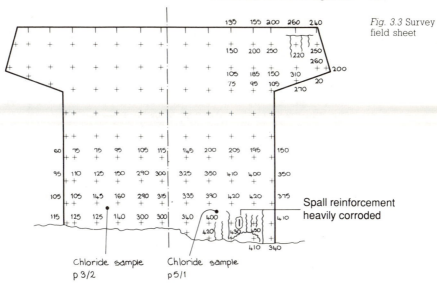

Fig. 3.3 Survey field sheet

Spall reinforcement
heavily corroded

Chloride sample
p 3/2

Chloride sample
p 5/1

Pier Three

Date: 2 Feb 1991   Face South

Bartown bridge survey

Category B
Cracks less than 0.5 mm wide.
Minor surface defects, honeycombing, surface staining or efflorescence
but with no sign of on-going deterioration.
Any previous repairs in good condition.

Category C
Cracks greater than 0.5 mm wide.
Local spalling and lack of adequate cover with signs of reinforcement
corrosion.
Extensive honeycombing or other surface defects.
Any previous repairs, cracking or spalling.

Category D
Visible distress to member: deflection, crushing, etc.
Significant widespread corrosion or spalling.
Doubt about the structural integrity of the member.

These are obviously very broad classifications. Their principal use is
in preliminary inspections of a group of structures so that they can be
ranked in terms of degree of deterioration. This information, along with
consideration of the strategic importance of the structures and the

consequences of failure, allows rational decisions to be made on the order in which repairs and refurbishment should be carried out.

Some owners of a large number of structures in the United Kingdom and elsewhere have developed their own categorization systems. The Department of Transport in the United Kingdom[3.11] requires that, for bridges, the severity and extent of defects are each assessed against a four point descriptive scale. These are as follows:

Severity of defects
1. No significant defects.
2. Minor defects of a non-urgent nature.
3. Defects of an unacceptable nature which should be included within the next two annual maintenance programmes.
4. Severe defects where action is needed within the next financial year.

Extent of defects
1. No significant defect.
2. Slight, not more than 5 per cent of area affected.
3. Moderate, 5−20 per cent of area affected.
4. Extensive, over 20 per cent of area affected.

Similar categories using the classifications Good, Fair, Poor, which equate to severities 1, 2 and 3 above, are used in conjunction with a standard inspection report form.[3.12] These categories are used for all elements of a bridge listed on the form, including non-concrete items such as fenders, bearings and metal deck plates. A similar good/fair/poor classification is used by British Rail for surveys of their structures.[3.13]

Fire-damaged Structures

A classification system for fire-damaged structures has been drawn up by the Concrete Society in the United Kingdom.[3.14] In the system there are five classes of damage for columns, walls, floors and beams. The classification is given in tabular form in the report but can be described generally as follows:

Class 0 Unaffected or outside the region of the fire.
Class 1 Some peeling of plaster or other finishes. Concrete has normal colour and no cracks, deflection or distortion with only slight crazing and minor spalling. No main reinforcement is exposed, except for beams which may have very minor exposure.
Class 2 Substantial loss of plaster or other finishes. Surface colour of concrete may be pink depending on aggregate type. No cracks, deflection or distortion with moderate crazing and localized spalling. Up to 25 per cent of main reinforcement is exposed

without buckling on beams and columns or up to 10 per cent exposed, but still adhering on walls and floors.

Class 3 Total loss of plaster or other finishes. Surface colour of concrete is whitish grey. Minor or small cracks present with no distortion or deflection of columns and no significant distortion or deflection of other members. Considerable spalling with up to 50 per cent main reinforcement exposed but still generally adhering on walls and floors. Only one of the exposed bars buckled for each beam or column.

Class 4 Plaster or other finishes destroyed. Concrete surface is buff coloured. Major cracks present on columns; severe and significant cracks on other members. Distortion of columns; severe and significant distortion or deflection of other members. Almost all of the surface is spalled with over 50 per cent of main reinforcement exposed on beams and columns and over 20 per cent exposed with much separated from the concrete on walls and floors. More than one bar buckled on columns and beams.

The system is very easy to use. A simple diagrammatic plan is drawn up for each floor showing the columns and the beams as straight lines. Each column, beam and floor panel is given an identification number. These plans are marked up on site to show the damage classification for each member as shown in Fig. 3.4. A second table in the report suggests appropriate repair techniques for each of the damage classes.

Although the Concrete Society system was developed specifically for the case of fire damage, it is relatively easy to conceive similar systems for other forms of damage. When developing a classification system the future rehabilitation of the structure should be kept in mind. A system which can be translated directly into an appropriate repair type or technique is very useful as it reduces the need for further surveys but this may not be possible in every case.

Commissioning Testing Work

It may be necessary to involve a specialized investigating and testing company in survey work. This will usually be after an initial survey has been carried out and when some impressions as to the cause of deterioration have been formed. When the use of a specialized company is considered, a full specification for the work should be drawn up. This should state the objectives of the survey and indicate the locations to be tested.

A description of the type of tests required should be given making reference to applicable national standards. The frequency of testing should

Fig. 3.4 Fire
damage survey
field sheet

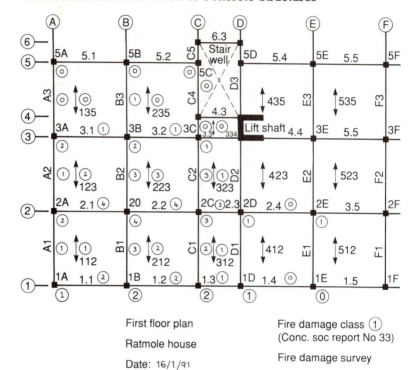

First floor plan

Ratmole house

Date: 16/1/91

Fire damage class ①
(Conc. soc report No 33)

Fire damage survey

be stated for equipment such as the half-cell or cover meter. For half-cell work this can be indicated by reference to a suitable grid spacing. For cover meter work various approaches to specification are appropriate depending on whether a structural assessment is being carried out. Typical areas of a stated size could be chosen in which every bar in both directions would be located and the cover measured. Alternatively line traverses could be specified at say 1 or 2 m intervals. The lines would be chosen at right angles to the direction of the bars with the least cover, and the spacing and the cover to the reinforcement determined.

References

3.1   ACI Committee 437 1985 *Strength Evaluation of Existing Concrete Buildings* American Concrete Institute, Detroit

3.2   Menzies J B 1978 Load testing of concrete building structures. *The Structural Engineer* **54A (12):** 347–53

3.3   Health and Safety Executive 1990 *Evaluation and Inspection of Buildings and Structures* Health and Safety series booklet HS(G)58, Her Majesty's Stationery Office, London

3.4   Health and Safety Executive 1980 *Entry Into Confined Spaces* Guidance Note GS5, Her Majesty's Stationery Office, London

3.5   Fookes P G, Pollock D J, Kay E A 1982 *Concrete in the Middle East Part 2* Viewpoint Publications, London

3.6   Institution of Structural Engineers 1980 *Appraisal of Existing Structures* Institution of Structural Engineers, London

3.7   Dixon D E 1986 Sampling for attributes. *Concrete International* **March:** 53–5

3.8   Davies S G 1976 *Further Investigations into the Strength of Concrete in Structures* Technical Report 41.514, The Cement and Concrete Association, Slough

3.9   Fookes P G 1976 A plain man's guide to cracking in the Middle East. *Concrete* **September:** 20–2

3.10   ACI Committee 201.IR 1984 *Guide for Making a Condition Survey of Concrete in Service* American Concrete Institute, Detroit

3.11   Department of Transport 1983 *Bridge Inspection Guide* Her Majesty's Stationery Office, London

3.12   Department of Transport 1988 *Trunk Road and Motorway Structures — Records and Inspection* Trunk Road Management and Maintenance Notice 2/88, The Department of Transport, London

3.13   Sowden A M (ed) 1990 *The Maintenance of Brick and Stone Masonry Structures* E & F N Spon, London

3.14   Concrete Society 1990 *Assessment and Repair of Fire-damaged Concrete Structures* Technical Report No 33 The Concrete Society, London

4 Testing Techniques

The types of tests carried out in-place on a structure or in the laboratory on recovered samples will depend on the underlying reason for the survey and on other criteria, such as access and the time available. If the objective is to estimate load-carrying capacity the tests will be aimed at assessing the strength of the concrete and the position, size and condition of the reinforcement. If the objective is to determine a cause of deterioration, strength may not be of primary interest, and chemical and other tests on both concrete and steel will be required.

Strength Testing

It is generally recognized that the concrete in a structure cannot be expected to have the same strength as the concrete in the test cubes or cylinders which are made for control purposes at the time of construction. If the procedures are carried out according to specification, the concrete in the control specimens is compacted in a standard way and the specimens are stored under water or at 100 per cent relative humidity and kept at a controlled temperature. The concrete in the structure may not be compacted to the same degree and it may lose moisture rapidly after placing. In the centre of large pours the concrete may reach high temperatures. In winter, freshly placed concrete may be exposed to low temperature which affects the rate of strength gain. In the longer term, the concrete in a structure may be subjected to physical and chemical processes which could significantly change its strength. Examples of this are conversion of high alumina cement concrete and sulphate attack on foundations.

There are now many techniques available for assessing the compressive strength of concrete in structures. The most direct method is the cutting and crushing of cores. This technique may not be appropriate if strength is to be determined at a large number of locations. In this case, a range of non-destructive tests is available to measure other related properties in-place on the structure. Compressive strength can be assessed if its relationship to the measured property is known or can be established.

A list of some non-destructive tests from which strength can be assessed has been given in Table 3.1 in the previous chapter together with brief comments on the accuracy of the resulting strength indication, ease of use and any resulting local damage to the structure. The individual tests are discussed in more detail in later sections of this chapter.

Core Testing

A core is a cylinder of concrete cut using a hollow drill barrel. The cutting edge of the barrel may be tipped with industrial diamonds and is usually cooled and flushed by water supplied through a swivel arrangement. The drill may be powered by electricity, petrol or by pneumatic means. Smooth-sided, straight cores are necessary for strength testing and these can only be obtained if the coring machine is anchored firmly in position. This can be achieved by bracing, anchor bolting, heavy weights or vacuum pads which can be used on smooth surfaces. It is not always necessary to have a piped water supply at the site. Cooling water can be supplied from a drum or tank situated above the cutting edge. Cores can be drilled upwards into slabs but the disadvantage is that cooling water runs back down the barrel towards the drill so that electrically driven machines are not suitable for this situation.

It is preferable that the cores do not include reinforcement as this can affect crushing strength, and it is a wise precaution to undertake a cover-meter survey before drilling to identify suitable locations. Normally, 150 mm or 100 mm diameter cores are employed for strength tests. For highly congested areas it may be thought necessary to use smaller diameters so that reinforcement can be avoided.

Lengths of the prepared samples for testing are usually between one and two times the diameter. Short cores are more often used than long cores, with length to diameter ratios in the range 1.0−1.2 being common. Short cores are preferable:

1. On cost grounds (drilling contractors often charge by length of core).
2. Because testing machine platen spacing is close to that required for standard cubes (150 mm). If the test is carried out on a machine used for cubes, it should perform satisfactorily for cores.
3. Long cores have a greater chance of containing an unrepresentative low strength area, which will result in a lower crushing value.

Small diameter cores, e.g. 50 mm, can be used satisfactorily for strength testing. Research has shown[4.1] that the mean strength obtained using small diameter, as an alternative to normal diameter, cores will be similar, but that there will be a greater spread in the results from small cores. It is therefore necessary to take a greater number of cores of small diameters.

When the core has been extracted it should be indelibly marked with a reference number, wrapped in cling film and sealed in an air-tight plastic bag so that it reaches the laboratory in an uncontaminated condition. In some circumstances it may be necessary to mark the cast face of the core so that its orientation can be determined. A record should be kept including the location in the structure from which the core was cut, whether the axis of the core was vertical or horizontal, and the date of extraction.

The British standard[4.2] on testing cores for compressive strength requires that each core is examined and described under several headings. Under the heading 'compaction' each core is examined for the presence of voids, honeycombing and cracks. The compaction can be further described by comparing the surface of the core with four photographs which illustrate what is termed 'excess voidage' in the range 0−13.0 per cent. Excess voidage is the amount by which the actual voidage exceeds the voidage of a well made cube. The standard defines voids in three size ranges. A small void is defined as measuring not less than 0.5 mm and not more than 3 mm in any direction. Medium voids have a maximum dimension between 3 mm and 6 mm, and large voids have a dimension greater than 6 mm. Further descriptions of the core may be given under the headings 'description of aggregate' and 'distribution of materials'.

Before testing the core, its density is determined. The density can be a useful guide to the condition and quality of the concrete, particularly its compaction.

The preferred method of end preparation in the British standard is grinding but capping with high alumina cement or a sulphur/siliceous sand/carbon mixture is also permitted. Cores are stored in water for half an hour prior to testing. The length to diameter ratio of prepared cores should preferably be in the range 1.0−1.2, but cores with a length to diameter ratio up to 2.0 can be tested under the standard.

Estimated *in situ* cube strength is calculated from the core crushing strength using the formula

$$\text{Estimated } in \; situ \text{ cube strength} = \frac{F}{1.5 + (D/L)} \qquad [4.1]$$

where F is 2.3 for vertical cores (in the member) or 2.5 for horizontal cores

D is the diameter of the core; and

L is the length of the core after capping.

Equations are also given in the British standard for correcting strength results for the presence of reinforcement in the core.

The procedures in ASTM C 42[4.3] are generally similar to those in the British standard. The ends of the specimen may be tooled or sawn and

Table 4.1 Strength correction factors for cores with different
length/diameter ratios from ASTM C 42[4.3]

Ratio length/diameter	Strength correction factor
1.75	0.98
1.50	0.96
1.25	0.93
1.00	0.87

capping methods are described in another standard[4.4] which permits the
use of high strength gypsum plaster or a sulphur mortar. Cores are stored
in lime-saturated water for at least 40 hours immediately prior to testing.
However, this soaking requirement is modified by ACI[4.5] procedures
which state that if the concrete is in a dry condition in service it should
be tested dry. The compressive strength results from cores with a capped
length to diameter ratio close to 2 (i.e. in the range 1.94−2.10) are
reported as the test value, but correction factors are applied to results
from cores with other length to diameter ratios as shown in Table 4.1.

The Rebound Hammer

This is one of the oldest non-destructive methods of assessing variation
of concrete strength within a structure. The equipment was developed
by Ernst Schmidt in Switzerland in the 1940s and in some countries the
rebound hammer is still known as the Schmidt hammer. The hammer
works by impacting a spring-loaded mass on a plunger which is in contact
with the surface. The distance which the mass rebounds is a measure
of the hardness of the surface.

The principle of operation is shown in Fig. 4.1. To perform the test,
the plunger, in its fully extended position, is placed against the concrete
surface. The hammer body is pressed firmly towards the concrete. This
causes the plunger and mass to slide towards the back of the hammer
extending the spring. As the plunger reaches the limit of its travel the
release catch is activated. The force in the spring causes the hammer
mass to travel along the hammer guide and to strike the shoulders of
the plunger. The hammer mass rebounds and engages an indicator on
a sliding scale which records the rebound distance.

The use of the rebound hammer is covered by BS 1881: Part 202[4.6]
and ASTM C 805[4.7]. ASTM C 805 states that the rebound method is
not intended as an alternative to strength determinations, but that it may
be used to assess uniformity of concrete, to determine areas of poor
quality or deteriorated concrete, and to indicate changes in characteristics
with time. The area to be tested should be at least 150 mm in diameter.
Shuttered surfaces or smooth trowelled surfaces can be tested without

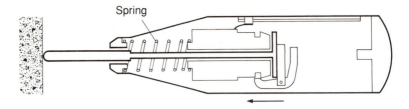

Fig. 4.1 Schmidt rebound hammer

Spring

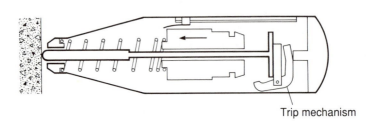

Trip mechanism

Sliding scale

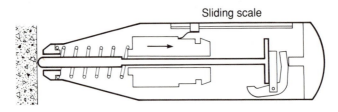

preparation but heavily textured surfaces or soft surfaces have to be ground smooth. Differences in moisture content can be reduced if the test area is thoroughly soaked for 24 hours prior to testing. Ten readings are taken from each test area with no two tests closer together than 25 mm. Immediately after each reading the impression made by the hammer on the concrete surface is examined. If the impact has crushed the mortar covering a void near the surface the reading is disregarded.

The average of the ten readings is calculated. If any reading differs from the average by more than seven units, that reading is discarded and the average recalculated. The entire set of readings is discarded if more than two readings differ from the average by more than seven units.

BS 1881: Part 202 also states that the use of a rebound hammer is not generally considered to be a substitute for other methods of determining strength. It is stated to be useful only as a preliminary or complementary method. The standard does, however, describe a method for obtaining a correlation between strength and rebound number. The test procedure is similar to that described in the ASTM. For a particular series of tests, the moisture condition of the surfaces during testing should be similar. Dry surfaces are preferable. The standard suggests that twelve readings are taken from an area 300 mm^2 and also that the readings are taken on a regularly spaced grid to reduce operator bias. The suggested spacing is 20−50 mm. All readings taken at a particular location are used in the calculation of the mean, unless there is a good and obvious reason for discarding a particular value.

The rebound hammer provides an empirical measure of the hardness of a localized area of the concrete surface. Part of the energy of the hammer mass is absorbed by the action of the plunger on the concrete. There may be local crushing of the surface and a stress wave is propagated through the member. The amount of energy absorbed depends on the stress−strain relationship of the concrete or its stiffness. A low strength concrete with low stiffness will absorb more energy than a stronger concrete with greater stiffness. However, there is no unique relationship between rebound number and strength. If the rebound hammer is to be used to assess strength then it is necessary to determine an individual correlation curve for the concrete under consideration. This is done by cutting cores from the structure at locations which have been previously tested with the rebound hammer. This procedure may not give results over a sufficient range to produce an adequate correlation. An alternative is to make cores from mixes using similar cement and aggregates to those in the structure under test. The mix proportions are varied to match the range of strengths likely to be encountered in place.

To test the cubes with the rebound hammer they are clamped in position in the crushing machine. Nine readings are taken on each of the two faces which are accessible as the cube is clamped in the machine. The test positions should not be within 20 mm of one another or within 20 mm of an edge. The mean of the rebound numbers and the crushing strength are used to construct the correlation curve.

ACI Committee 228[4.8] has described a similar method of correlating rebound number with strength using cylinders. A range of strengths is achieved by testing specimens from the same mix at different ages. It is suggested that the testing should be arranged so that a minimum of six strengths at approximately even intervals is obtained. Two cylinders are tested at each age and ten rebound readings are obtained from each cylinder as they are held firmly in a compression testing machine. It is noted that accuracy is increased if the moisture condition of the correlation specimens is similar to that in the structure and that saturation is the only easily reproducible moisture condition.

There are many factors which affect the results of the rebound test. Firstly the test is highly localized and only the concrete within approximately 30 mm of the surface contributes to the result. It is well known that this surface layer may not be representative of the concrete at depth. The surface layer may be carbonated, which tends to increase hardness or it may also be less well cured because of rapid loss of moisture from this zone. In a similar way, the hardness may vary across a structure because of changing exposure conditions. The type of formwork can affect the rebound number and trowelled surfaces generally give higher results than formed surfaces.

The rebound hammer can be used on surfaces of any orientation including soffits. Clearly the design of the rebound hammer means that different conversion factors will apply depending on the inclination of the axis of the hammer.

Ultrasonic Pulse Velocity

Ultrasonic methods have been used for assessing the strength of concrete, detecting flaws such as voids and cracks, and estimating thickness of layers where the layers have differing sound propagating properties. The method can, therefore, be used to estimate the approximate depth of damage due to such causes as frost, sulphate attack or fire.

The velocity of a sound compression wave in any uniform elastic medium is governed by the equation:

$$V = (K \times E/d)^{0.5} \tag{4.2}$$

Where V is the wave velocity in km s^{-1};
  K is a function of Poisson's ratio of the medium;
  E is the dynamic modulus of elasticity in kN mm^{-2}; and
  d is the density of the medium in kg m^{-3}.

For a particular concrete mix the elastic modulus varies with the square root of strength if other factors remain the same. Hence, measurement of the velocity of sound provides a means of estimating strength for comparative purposes.

Most commercially available equipment uses pulses of ultrasound and the basic layout is shown in Fig. 4.2. A pulse generator causes repeated voltage pulses to be transmitted to a transducer which is coupled to the concrete surface. Similar pulses are sent to a timing circuit. The transmitting transducer causes compression waves to be generated in the concrete. These are picked up by a similar receiving transducer, which converts the mechanical energy of the compression waves back into pulses of electrical voltage which are sent through an amplifier to the timing circuit. The transit time is displayed in digital form in microseconds. If the distance between the transducers is known the pulse velocity can be calculated.

The time measured by the instrument includes the time taken for the voltage pulse to travel along the cables to and from the transducers and also the time taken for the ultrasound pulse to travel through the transducers. The instrument usually includes, therefore, a time adjustment

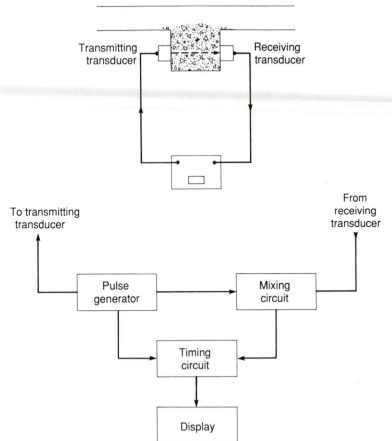

Fig. 4.2 Ultrasonic pulse velocity equipment (after Bungey, J H, *The Testing of Concrete in Structures*)

control which can be used to eliminate these errors. To make the adjustment, the transducers are coupled to the opposite ends of a cylindrical bar for which the transit time is known accurately. The instrument is then calibrated by adjusting to show the known transit time. Reference bars are usually made from steel and are supplied with the instrument. The adjustment needs to be made on each occasion on which the equipment is used, when transducers or cables are changed and also from time to time during operation of the equipment.

Transducers are usually in the form of short metallic cylinders which have to be carefully coupled to the concrete surface. Other types of transducer are available as discussed below. When the concrete surface is relatively smooth, this can be achieved with liquid soap or grease. Only a thin layer must be used and the transducer should be pressed firmly against the surface to avoid air pockets and surplus couplant. On rough surfaces or in hot climates a stiffer grease may have to be used. The grease on the transducers tends to pick up dust and grit, and the transducers should be thoroughly cleaned each time they are relocated. On very rough surfaces it may be necessary to grind smooth at the location of transducer positions or to provide small pads of plaster of Paris or a quick-setting mortar.

Most of the difficulties normally encountered in use of pulse velocity equipment are associated with poor coupling. Normal good procedure is to monitor the readings as they are produced and repeat any measurements that are considered doubtful. Repeat readings should be undertaken by removing both transducers, cleaning the transducers and the surface of the concrete, re-applying couplant and firmly pressing the transducers onto the surface. This process may have to be repeated until a minimum constant reading is obtained.

Exponential probe transducers are also available. These have a very small contact area with the concrete surface and are easier to use on curved or rough surfaces, and on flat smooth surfaces may allow substantially more readings to be obtained in a given period of time.

The use of ultrasonic pulse velocity measurement equipment on concrete is covered by BS 1881: Part 203[4.9] and ASTM C 597.[4.10] The ASTM states that measurement of pulse velocity should not be considered as a means of measuring strength or for establishing elastic modulus. The main uses described in the standard are:

1. Assessing uniformity of concrete in the field.
2. Indicating changes in the characteristics of concrete with time.
3. Assessing the degree of deterioration and/or cracking when surveying concrete structures.

However, in a note it is explained that a velocity−strength relationship can be established by conducting pulse velocity and compressive strength tests on the same samples of a particular concrete, and that, if a good correlation is found this may serve as an index for estimating strength on the basis of further pulse velocity tests on the same concrete.

The ASTM contains sections on procedures for carrying out measurements and also procedures for determining calibration correction and zero correction by pressing together the transducers. BS 1881: Part 203 lists similar applications for pulse velocity measurement to those given in ASTM C 597. However, it also suggests correlation of pulse velocity and strength as a measure of quality and sets out a suitable procedure for obtaining a correlation.

The British standard describes three different transducer configurations which may be employed. These are known as direct, semi-direct and indirect (Fig. 4.3). In the direct configuration, the transducers are placed on opposite faces of the member under consideration and the ultrasonic

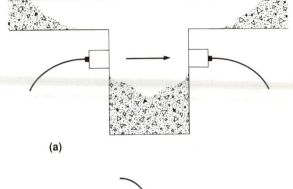

(a)

Fig. 4.3 Ultrasonic
pulse velocity
equipment
transducer
configurations.
(a) direct
configuration;
(b) semi-direct
configuration;
(c) indirect
configuration

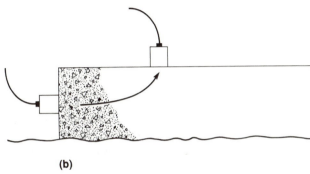

(b)

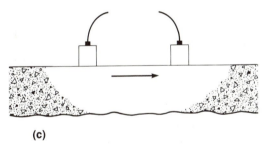

(c)

pulse travels directly through the member. In the semi-direct configuration, the transducers are placed on adjacent faces of the member or on opposite faces, but with the transducers not directly opposite one another. The pulse again travels through the body of the concrete. In the indirect method, both transducers are placed on the same surface of a member (e.g. a ground slab). In this case the pulse travels through the concrete in a region just below the surface.

The direct method allows the maximum transfer of energy between transducers and the most accurate measurement of pulse velocity. This configuration should be used whenever circumstances permit. The indirect configuration leads to only a small percentage of the transmitted energy arriving at the receiving transducer and is, therefore, relatively insensitive.

BS 1881: Part 203 suggests that, when using the indirect configuration, a series of results should be taken at each location with the transducers at different distances apart. A graph is then plotted showing the relationship between transmission time and the distance between the transducers. The slope of the straight line graph drawn through the points is taken as the mean pulse velocity along the line of measurement on the concrete surface.

There is no unique relationship between ultrasonic pulse velocity and strength of concrete. The relationship will depend on the type of aggregate and cement used in the production of the concrete and the mix proportions. For a particular mix, the relationship is affected by age, conditions during curing and the local moisture content. If the technique is to be used to give an estimate of strength in a structure, it is necessary to establish the correlation for the particular concrete under test. BS 1881: Part 203 describes two methods of establishing a correlation. In one method, cube, beam or cylinder specimens are produced using similar ingredients to those in the concrete under consideration. The strength of these specimens is varied by changing the water/cement ratio or testing at different ages. At the time of test the specimens are tested for both ultrasonic pulse velocity and strength. Three specimens are tested from each batch or at each age, and at least three measurements taken at different heights on each specimen. The mean pulse velocity and mean strength are used to construct the correlation curve.

The second method described in BS 1881: Part 203, using cores cut from the structure, is more appropriate in the case where the strength of concrete in a structure is being investigated many years after construction, as the details of the concrete mix are rarely known with certainty and the correlation between strength and pulse velocity changes with the age of the concrete. The pulse velocity technique is first used on the structure to locate areas of concrete with different pulse velocity. Locations are then chosen for coring on the basis of the readings to give a wide range of strengths. The correlation is based on the strength of the cores and the pulse velocity through the concrete at the core locations. It is likely that only a limited range of core strengths and pulse velocities will be obtained for the construction of the curve. However, BS 1881: Part 203 points out that the correlation can be extrapolated as the shape of the curve is the same for all curing conditions.

It may be possible to determine the depth of cracks and the thickness of surface layers using the indirect configuration. The method employed is to take readings with the transducers at different distances apart. In the case of a crack perpendicular to the surface, the transducers are placed

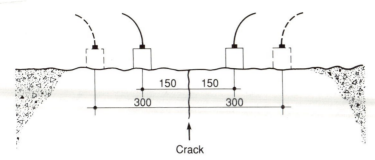

Fig. 4.4 Layout for
crack depth
assessment (after
Bungey)

150   150

300   300

Crack

at two convenient distances equally spaced on opposite sides of the crack
as shown in Fig. 4.4. The pulse cannot travel directly between the
transducers along the surface because of the presence of the crack.
However, some of the energy from the transmitter will travel around
the tip of the crack and to the receiver. If the transducer and receiver
are placed at distances of 150 mm and 300 mm from the crack, the crack
depth is given by:

$$c = 150 \times \sqrt{\{(4t_1^2 - t_2^2)/(t_2^2 - t_1)\}} \qquad [4.3]$$

where t_1 is the measured transit time in microseconds when the
transmitter and receiver are each 150 mm from the crack; and
t_2 is the measured transit time in microseconds when the
transmitter and receiver are each 300 mm from the crack.

In the case of the measurement of the thickness of a surface layer,
a graph of transit time against transducer spacing is plotted as shown
in Fig. 4.5. When the transducers are close together the measured pulse
travels mainly through the surface layer. As the spacing between the
transducers increases there will come a time when the shortest transit
time will be achieved by travel along the surface of the underlying layer.
If the velocity of the pulse in the two layers is different, there will be
a change in slope of the plot of transit time against spacing.

If s is the spacing at which the change of slope occurs, the thickness
of the surface layer is calculated from:

$$t = s/2 \times \sqrt{\{(V_u - V_s)/(V_u + V_s)\}} \qquad [4.4]$$

where t is the thickness of the surface layer in millimetres;
s is the transducer spacing at which the change of slope occurs;
V_u is the pulse velocity in the underlying layer in km s^{-1} (the
slope of the graph when the transducer spacing is greater than
s); and
V_s is the pulse velocity in the surface layer in km s^{-1} (the slope
of the graph when the transducer spacing is less than s).

Fig. 4.5 Layout for
assessment of
depth of a surface
layer and a plot of
results (after
Bungey)

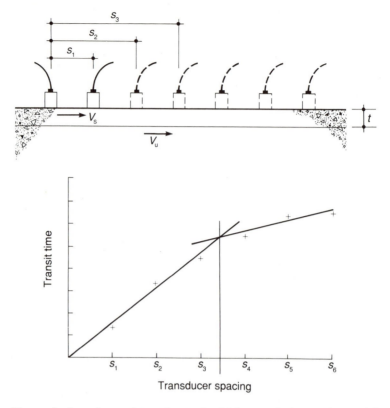

The method can be used to estimate the thickness of a layer of materials
with different properties so long as the velocity in the surface layer is
lower than that in the underlying layer.

There are several factors which can influence pulse velocity
measurements. The principal factors are degree of compaction, moisture
content and the presence of reinforcement. In a poorly compacted
concrete, numerous small voids lower the density and strength. They
also cause the ultrasonic pulse to travel over a longer path and give a
lower indicated velocity. Unless the reductions in strength and velocity
are matched, erroneous estimates of strength can result. A 5 per cent
change in pulse velocity results in a change in estimated strength of the
order of $10 \, \text{N} \, \text{mm}^{-2}$ for a typical correlation.

Penetration Resistance

The penetration resistance method of assessing strength uses a hardened
steel probe which is fired into a concrete surface using a special gun
and a standardized explosive charge. The surface is cleared of loose
material and a steel plate is positioned adjacent to the probe. A steel

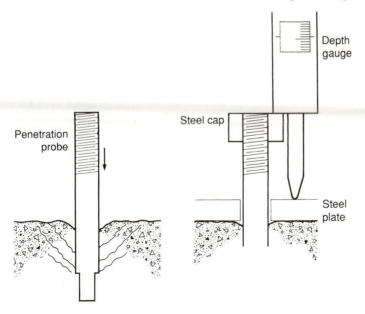

Fig. 4.6
Penetration
resistance test

cap is screwed onto the probe and the length of probe protruding from the surface is measured using a spring-loaded depth gauge (Fig. 4.6) and the depth of penetration can be determined.

It is clear that the forces acting on the probe and the processes occurring within the concrete are extremely complex. As the tip of the probe enters the concrete, it causes local crushing of the surface and creates a shock wave which results in local spalling. As penetration continues, the forward motion of the probe is resisted by friction along its sides, concrete is crushed at the tip and the concrete ahead of the probe is compressed in a zone of indeterminate shape and size. The initial kinetic energy of the probe is dissipated by all of these processes and it is asserted that the depth of penetration can be calibrated against compressive strength. Cartridges are available at a choice of two power levels according to the strength of the concrete to be tested.

The use of the penetration resistance method is described in ASTM C 803[4.11]. The standard notes that penetration resistance testing can be used to obtain an indication of the development of strength in concrete, to investigate areas of deterioration or poor quality, and to assess the uniformity of concrete within a structure, but that the method is not intended to take the place of strength determination by conventional methods. The standard also notes that a relationship may be established between penetration resistance and strength for a particular concrete mix and a particular test equipment and used to assess strength in structures. However, the relationship may be affected by curing conditions and the type and size of aggregate.

ASTM C 803 requires that rough concrete surfaces are ground smooth in the test area prior to test. Grinding is required if the surface is coarser than that produced by a burlap (hessian) dragged finish. Tests are not to be carried out within 100 mm of an edge or 175 mm of a previous probe. At least three probes are to be employed at each location and each probe is checked by tapping with a small hammer to make sure that it has not rebounded from the position of maximum penetration, and to confirm that it is embedded firmly. Results from any loose probes are rejected.

There is, at present, no individual British standard which deals with the penetration resistance test method, but it is understood that it will be covered in BS 1881: Part 207[4.12] to be published at some time in the future. Some brief details of the use of the test are given in BS 1881: Part 201[4.13]. This standard states that it is normal to carry out three tests at each location and take the mean as the result. A minimum edge distance and member thickness of 150 mm are suggested. It is noted that calibration for the particular aggregate in use is essential if compressive strength is being assessed, and also that estimated strength values are unlikely to have an accuracy of less than ± 20 per cent at confidence levels of 95 per cent.

It is more inconvenient to produce calibration curves for the penetration resistance method than for the ultrasonic pulse velocity method or the rebound hammer method. Because of the minimum edge distance requirements the calibration cannot be carried out on standard cubes or cylinders. It may be possible to use standard specimens for lower strength concretes using the lower powered cartridges, but it is necessary to produce pairs of samples as probing would reduce the crushing strength if the same sample were used for both penetration resistance and crushing. For concrete of higher strength, it is necessary to use standard compression strength specimens and larger specimens compacted and cured under similar conditions from the same batch for penetration resistance.

One of the virtues of the penetration resistance test is its ease and speed of operation. It is also said to be relatively independent of operator technique and insensitive to factors such as moisture content. Among the disadvantages are the relatively high cost of probes, which are used only once, and the fact that damage is caused to the surface which then has to be repaired. There is also the possibility that results could be affected by the presence of particularly hard or soft aggregate.

Pull-out and Pull-off Tests

Several tests are now available which give estimates of compressive strength by measuring the force required to pull out embedded anchors or to pull off discs glued to the surface. Many of the pull-out tests use anchors which are placed in the concrete at the time of construction. Their main use is for testing strength development in new concrete in precast works and in research, and they are not suited to the investigation of structures. However, some pull-out tests use wedge anchor bolts which are inserted into holes drilled into the concrete and these tests can be used.

BRE Internal Fracture Test

The Building Research Establishment (BRE) internal fracture test[4.14] is the pull-out test which has been most widely used in the field. It was first developed for testing high alumina cement concrete components but further work has been carried out to produce correlation curves for Portland cement concretes. To carry out the test a 6 mm hole is drilled to a depth of 30−35 mm and the hole is blown out. Holes for individual tests should be at least 100 mm apart, should not be close to an edge and should be drilled away from reinforcement.

A depth mark is made on a standard wedge anchor bolt at a position 20 mm above the lower edge of the wedge sleeve. Adhesive tape is often used for this purpose. The bolt is tapped into the hole until the depth mark is at the concrete surface. The threads on the bolt are greased so that the nut travels freely. A reaction frame with a spherical seating is placed over the bolt so that it bears on the concrete surface, as shown in Fig. 4.7. If necessary, washers are placed on the reaction frame so that only approximately 5 mm of the bolt projects above the washers.

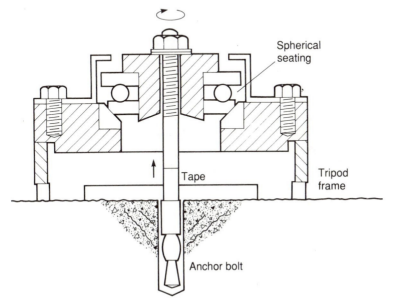

Fig. 4.7 BRE pull-out test

This is to make it less likely that the nut reaches the end of the threaded portion of the bolt before completion of the test. The nut is tightened on the bolt with the fingers to settle the thrust pad on its spherical seating. A torque meter is used to tighten the nut until a reading of 1 N m is indicated. The nut is tightened half a turn at a time, using the torque meter, smoothly and without jerking. Each half turn should take approximately ten seconds. The torque is recorded at each half turn and the procedure is repeated until the torque reaches a maximum and then begins to reduce.

It is usual to carry out six tests on a member. The average of the six maximum readings is used to read off the mean compressive strength from a standard calibration curve given by:

$$f = 3.116T^{1.69} \hspace{4cm} [4.5]$$

where f is the cube crushing strength in $N\,mm^{-2}$; and
T is the average maximum torque in $N\,m$.

It is suggested in the Building Research Establishment Information Paper 22/80 that sets of results should be examined as work proceeds so that individual 'rogue' results much higher or lower than the average can be quickly identified. The standard calibration graph gives upper and lower 95 per cent confidence limits, and it is suggested that any readings which lie outside of these limits at the mean should be discarded. Where individual results are discarded, repeat tests should be carried out if possible so that the average is based on six results which lie within the 95 per cent confidence limits at the mean. It is suggested that the method allows concrete strength to be estimated to within ±30 per cent at the 95 per cent confidence level.

The test causes internal fracture of the concrete and there should be little damage to the surface. After the test the bolt can be sawn off and the surface made good.

The Capo Test

The Capo test is a modification of the Lok test which was developed in Denmark. The Lok test uses cast-in inserts and conforms to ASTM C 900.[4.15] The Capo test's name is based on the initial letters of the phrase 'cut and pull out'. To carry out the test, an 18 mm hole is drilled to a depth of 45 mm. The hole is reamed out to form a 25 mm diameter groove at a depth of 25 mm. A ring insert is placed in the hole and expanded in the groove as shown in Fig. 4.8. The insert is connected to the standard Lok test equipment, which consists of a hand-operated hydraulic jack and load-indicating gauge. Connection to the test equipment is made by a threaded coupling rod 7.2 mm in diameter. The hydraulic jack produces a tensile force in the rod and bears on a reaction ring of internal diameter 55 mm on the concrete surface.

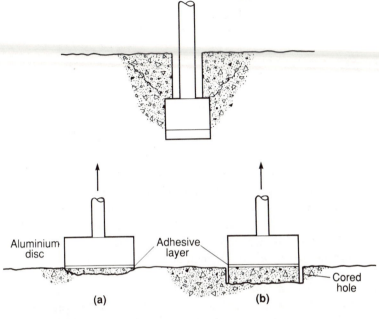

Fig. 4.8 The Capo test

Fig. 4.9 Pull-off test

The failure surface generated by the test is in the form of the frustum of a core from the outer edge of the expanding insert to the inner edge of the reaction ring. The test can be stopped and the load released as soon as the peak is reached. This results in only a fine crack on the surface. Alternatively, the test can be continued until a plug of concrete is removed so that the disc can be recovered and re-used.

Limpet Test

The Limpet test[4.16] is a pull-off test. An aluminium disc, 50 mm in diameter, is glued to the concrete surface at the test position using an epoxy resin or other suitable adhesive (Fig. 4.9). The disc is connected to a mechanical loading device by a threaded rod. Load is applied by turning a handle on the side of the machine. Applied load is measured by integral strain gauge equipment with digital read-out. The applied force pulls the disc away from the surface with an attached thin layer of concrete. The concrete fails principally in direct tension. It is normal to carry out three tests on a member and calculate a mean pull-off force. The compressive strength can be assessed using suitable calibration curves.

Since the layer of concrete detached with the disc is usually only a few millimetres thick, the results can be strongly influenced by surface effects such as laitance, poor curing or carbonation. This difficulty can

Fig. 4.10 Break-off
test

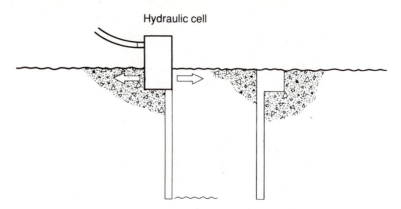

be overcome by core drilling using a 50 mm diameter barrel to a depth
of say 20 mm or greater, to penetrate concrete below the surface layer.
The central section is not broken off and the aluminium disc is attached
to it before pulling in the normal way. This technique is also useful for
testing the bond of repairs or overlays. In this case, the core is drilled
through the repair or overlay into the parent concrete.

Break-off Test

A test which measures the force required to break off a core has been
developed in Norway.[4.17,4.18] The test was originally developed to
monitor the early strength of concrete, and utilized a plastic cylindrical
insert at the time of casting to form the core. However, it can also be
used on drilled cores. A core of 55 mm diameter is drilled to a depth
of 70 mm. The annular hole on the surface is enlarged to form a circular
groove 10 mm wide and 10 mm deep to accept the loading device, as
shown in Fig. 4.10. The load is applied transversely at a depth of 5 mm
from the surface. The flexural tensile stress developed at the base of
the core causes it to rupture and the core breaks off.

Covermeter (Pachometer)

A covermeter is an electromagnetic device used for determining the
location and cover of reinforcing bars in concrete. The use of covermeters
is described in BS 1881: Part 204.[4.19] The instrument consists of a
search head connected by a cable to a metering unit which may have
a digital or analogue read-out.

Commercially available instruments work either on the principle of
magnetic induction or on the effects of eddy currents set up in the
reinforcement. In the former case, a coil carrying an alternating current
in the search head sets up a magnetic field which induces a current in
a second coil. If any steel object enters the magnetic field the induced

current is altered and the magnitude of the change in current depends on the size of the steel object and the distance from the search head. In the case of instruments using the eddy current principle, alternating currents in an activating coil set up eddy currents in the reinforcing bars. These eddy currents cause changes in the impedance of a measuring coil which are again related to the diameter of the bar and the distance from the search head. The eddy current technique can be used for detecting non-magnetic conducting materials and the calibration can be affected by change in steel type.

Part 204 of BS 1881 gives details of methods for carrying out laboratory checks on the calibration of a covermeter with a recommendation that they are carried out at least six-monthly intervals. Field calibration checks are carried out on each occasion the covermeter is used. Three methods of carrying out laboratory checks are described.

In the first method, a straight clean smooth reinforcing bar is cast into a concrete prism such that it is parallel to and has a different cover to each of four faces. The cover indicated by the covermeter can then be checked against directly measured values. The procedure can be repeated for bars of different diameter.

In the second method, the covermeter measuring head is placed on a table top remote from any metallic objects such as nails or screws. A straight clean smooth reinforcing bar is also placed on the table parallel to the search head. The distance between the head and the bar is measured and compared with the instrument reading. The procedure is repeated for the full depth range of the instrument and different bar diameters.

The third method uses a box constructed of non-metallic materials. Holes are drilled in two opposite sides of the box so that a straight clean smooth reinforcing bar can be fitted through them at different distances from and parallel to one face. The measuring head is placed on the surface parallel to the bar and the reading noted. The procedure is repeated with the bar at different distances from the measuring head.

In each of the above calibration methods the accuracy should be to within ± 5 per cent or ± 2 mm, whichever is the greater. This required accuracy applies over the manufacturer's full stated range and for all applicable bar diameters. Some constituents of concrete may affect covermeter readings and so it is always advisable to carry out a calibration check for the actual situation. This can be done by taking readings on bars at a range of different depths and either drilling down or breaking out to determine the actual cover. A calibration curve can then be constructed.

The zero reading of the instrument should be checked each time it is switched on and also at regular intervals during use. This is done by holding the search head remote from the concrete surface and any other metal object, and checking that the instrument is giving the appropriate reading. Most instruments operate on battery sources and include circuitry for checking the state of charge. Low charge is indicated automatically on some digital instruments, but if this is not the case it will be necessary to check the state of charge from time to time during operation.

The measuring head of the covermeter must be aligned in the correct direction in relation to the bar in order to give the most accurate cover reading. In most cases the directions of the bars will be reasonably easy to deduce. When faced by an irregular shaped member where the direction of reinforcement is uncertain it will be necessary to proceed as follows. Place the measuring head on the concrete surface with the head aligned in any direction. Move the measuring head, in contact with the surface, in the direction perpendicular to its longitudinal axis. The cover reading indicated by the instrument will decrease as the position of an underlying bar is approached. Adjust the measuring head until a minimum value is obtained. Gradually rotate the measuring head about its centre, still keeping it in contact with the surface. Adjust the angle of the head until a minimum reading is obtained. Mark the alignment of the head on the surface. Repeat the procedure starting with the measuring head at right angles to the direction previously determined. If the directions of the bars are determined at a number of locations a pattern should become apparent.

As stated above, the signal generated in the search head is dependent on bar size as well as the distance to the bar. Instruments with digital read-out allow for this by incorporating a range of settings for different bar diameters. Instruments with analogue read-out either have a number of different scales appropriate to different bar diameters or a single scale with broad divisions. For a particular cover, readings for small diameter bars will occur towards the deeper end of the division while readings for large diameter bars will occur towards the shallower end. With most covermeters it will be necessary to know the diameter of the reinforcement in order to obtain accurate readings. If reinforcing drawings exist for the structure under investigation the reinforcement size should still be checked by breaking out at a few locations. If drawings are not in existence a more comprehensive survey will have to be carried out. A method of assessing bar diameters using the covermeter is described below.

When measuring the cover on a member with reinforcement in two directions at right angles to each other such as a slab, it is probably best to quickly determine the locations of the bars by moving the head across the surface and marking the location of the minimum reading. Some digital covermeters omit an audio signal which assists this operation. The pitch of the signal increases as the bar position is approached. The measuring head can then be placed over the bars in the direction being measured and between the bars in the other direction to minimize their effect on the instrument reading. The same procedure can be adopted on beams and columns. Place the measuring head between the links or stirrups when measuring the cover to the main bars.

Some covermeters can also be used for assessing the diameter of reinforcing bars. These instruments have different scales for different bar diameters. The bar diameter is assessed by locating a bar using any arbitrary scale and then recording the indicated reading using each bar diameter scale. A spacer of known thickness, say 20 mm, is placed between the concrete surface and the measuring head and the reading using each scale is again recorded. The difference between the two readings on each scale is calculated. The diameter corresponding to the scale where the difference in reading is closest to the thickness of the spacer is taken as an indication of the underlying bar diameter.

Half-cell Potential

Corrosion is an electrochemical process as described in Chapter 2. The process of corrosion causes electrical potentials to be generated and the half-cell provides a method of detecting and categorizing these electrical potentials. The equipment and method are described in ASTM C 876.[4.20] There is no equivalent British standard but a brief description giving limitations and principal applications is contained in BS 1881: Part 201.[4.13] Drafts of an outline specification and a guide to potential mapping have been developed by the Institute of Corrosion in the United Kingdom.[4.21,4.22]

The half-cell consists of an electrode of a metal contained in an electrolyte consisting of a saturated solution of one of its own salts. Copper in copper sulphate and silver in silver chloride are commonly used. A porous plug in the end of the cell allows the electrolyte to make contact with the concrete surface. A diagram of a copper/copper sulphate cell is shown in Fig. 4.11. The half-cell provides a reference against which corrosion potentials of steel in concrete can be measured. As its name suggests, it can be thought of as being half of a cell or battery capable of developing a measurable potential. The other half of the cell is provided by the reinforcement when the connections are set up as shown in Fig. 4.12. The internal connection is through the electrolyte in the half-cell and the moisture within the concrete. The external connection is through a high-impedance meter which is capable of measuring the potentials generated.

Corrosion activity on the reinforcement will cause a change in the potential generated by the copper/steel or silver/steel cell. The half-cell is moved to different locations on the concrete surface, usually in a grid pattern, and the potential in millivolts at each location is noted.

For the method to work it is necessary that there is good electrical

Fig. 4.11
Copper/copper
sulphate half-cell

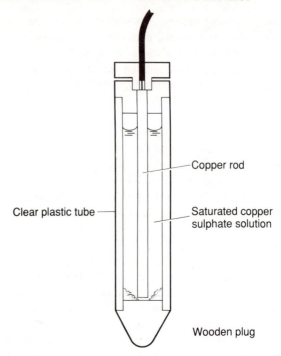

Copper rod

Clear plastic tube

Saturated copper
sulphate solution

Wooden plug

continuity through the reinforcement over the area being surveyed. The Institute of Corrosion draft outline specification suggests that the continuity should be checked by exposing the reinforcement at opposite corners of the area and measuring the resistance using a high-impedance voltmeter. The resistance should be less than 1 ohm.

It is also essential that a good contact is achieved with the exposed reinforcement and that the contact resistance between the half-cell and the concrete surface is minimized. The connection to the reinforcement can be made by drilling a hole and using a self-tapping screw, by brazing or by using a clamp or clip. If a clamp or clip is used, the surface of the reinforcement must be carefully cleaned at the location. Connections

Fig. 4.12 Half-cell
test layout

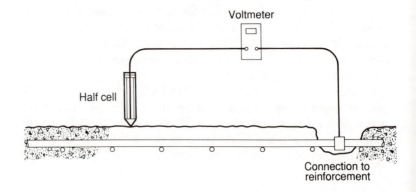

Voltmeter

Half cell

Connection to
reinforcement

with zinc or aluminium plating should be avoided as they could produce galvanic potentials which would alter the apparent half-cell readings. In order to minimize the contact resistance between the surface and the cell, the surface is cleaned and any surface coating is removed at the locations where the readings are to be taken. A liquid bridge between the cell and the concrete is also provided by either spraying the surface locally or by use of a sponge soaked with a conductive fluid. ASTM C 876 suggests a contact solution made up of potable water and a wetting agent (washing-up liquid) in the ratio of approximately 19 : 1. The Institute of Corrosion guide suggests either clean tap water or a weak sodium hydroxide solution.

Half-cells should be cleaned and calibrated regularly. In copper/copper sulphate cells the porous plug has to be kept in a moist condition otherwise the pores may become clogged with crystals; a rubber cap may be provided for this purpose. The solution should be changed regularly and the copper rod will occasionally need cleaning. This can be achieved by wiping with a dilute solution of hydrochloride acid or an abrasive nylon cloth. Clearly steel wool or other metallic materials should not be used for cleaning as contamination could affect the electrical readings. Cells should be calibrated before each day's use in the field. This can be done by checking either against a standard cell or a new cell of the same type.

The half-cell method lends itself readily to automatic data logging and some commercial systems are available. One system uses multiple cells mounted on a light bar. Another, designed for use on flat horizontal surfaces such as bridge decks, makes contact with the concrete surface through a wheel.

Resistivity

The rate at which the corrosion reaction can proceed is governed by the restivity of the concrete amongst other factors. *In situ* resistivity measurements are sometimes carried out in association with half-cell surveys. A four-probe Wenner array (Fig. 4.13) is often used for this purpose. Metal rods are inserted into four equally-spaced holes in a straight line within the concrete surface. An alternating voltage is connected to the outer two rods and the potential drop across the inner two rods is measured. The spacing between the rods is usually of the order of 50 mm. Resistivity is calculated from the equation:

$$P = 2\pi\delta E/I \qquad\qquad [4.6]$$

where P = resistivity in ohm cm;
    δ = probe spacing in cm;
    E = voltage drop across the two inner probes; and
    I = current flowing between the two outer probes.

Fig. 4.13 Resistivity
measurement

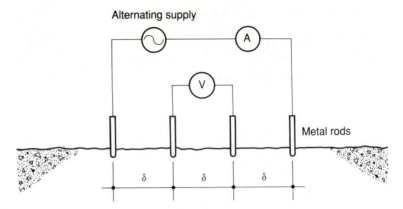

Radar

Ground-probing or subsurface interface profiling radar can be employed in a number of different ways in the investigation of concrete structures. The technique has been used for measuring the thickness of members, for determining the spacing and cover to reinforcement, and for detecting the position and extent of voids.

The technique is similar to the methods used in seismology except that radio frequency pulses are used rather than sound. The equipment generates electromagnetic pulses which are transmitted to the member under investigation by an antenna close to its surface. The pulses travel through the member but are partially reflected at any interfaces where there are distinct changes in dielectric characteristics. Reflected pulses are received by a second antenna and processed by the equipment.

The surface of the member is marked out with transit lines with regular distance markers. The antenna is moved along the transit lines and a signal is inserted into the system each time a distance marked is passed. Output from the instrument is in the form of a graphical time-based trace from an analogue recorder. The output shows, in the vertical direction, the time taken for the pulse to travel to an interface and return to the antenna, and therefore gives a rough profile of the underlying interfaces. The signals inserted at distance markers are also shown on the trace as a series of vertical lines. These allow the horizontal positions of any detected features to be determined.

Pulses tend to spread out from the transmitted point and therefore reflected signals are received from interfaces slightly ahead of and behind the position of the antenna. For this reason small reflectors such as reinforcing bars tend to show up as parabolic shapes on the output trace as shown in Fig. 4.14.

The radar signal is attenuated, or diminished in strength, as it passes

Idealized trace

Fig.4.14 Radar trace from reinforced slab

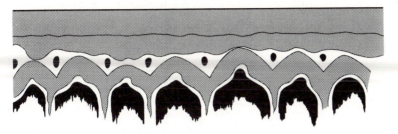

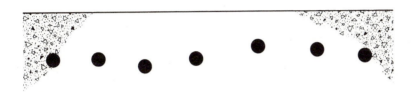

through the member under investigation. The amount of attenuation is dependent on the conductivity of the material and the frequency of the signal used. Attenuation is increased in conductive materials and therefore it may be difficult to penetrate saturated or salt-contaminated concrete. In general, low frequency radiation penetrates further than high frequency radiation. However, low frequency signals result in a loss of resolution, or ability to pick out small objects, because of the increase in wavelength.

The velocity of propagation of the radar pulse is dependent on the dielectric constant of the concrete. This can vary quite widely depending on aggregate type, moisture content and other factors. There is a need, therefore, for depths determined by radar to be checked and calibrated by direct physical measurements at a few locations.

Dust Sampling

Chemical tests on samples of concrete are required in the process of investigation of various forms of concrete deterioration. In many cases useful results can be obtained from quite small quantities of concrete (less than 50 g) which can be obtained quickly and cheaply by collecting drill dust. A hand-held rotary percussion drill can be used for this purpose. In order for the samples to be representative, a bit diameter of at least 10 mm is used. Various means of collecting the concrete dust can be employed. On vertical surfaces, a clean container such as a polythene sample bag can be held immediately below the drill hole or

Fig. 4.15 Methods of drill dust sampling. (a) Using cut plastic tube on horizontal surfaces; (b) using plastic cup on soffits

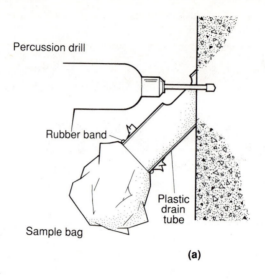

Percussion drill

Rubber band

Plastic drain tube

Sample bag

(a)

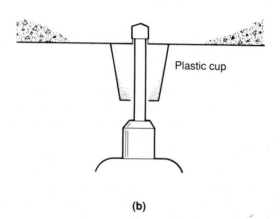

Plastic cup

(b)

group of drill holes. An alternative is to use a section of plastic tube cut off at an angle with a hole in the side as shown in Fig. 4.15(a). This method allows samples to be collected very quickly. Vacuum drills which extract the dust from the holes are available and are particularly useful on the top surface of slabs. On slab soffits, samples can be obtained by drilling through the base of a plastic cap as shown in Fig. 4.15(b).

In many cases, the variation of chemical properties with depth is of interest and it is necessary to obtain samples from different depths. In this case the drill hole or holes are advanced in stages and the dust from each depth increment is collected separately. The holes are cleaned out carefully between drilling to minimize cross-contamination of samples.

Carbonation Depth Testing

Depth of carbonation can be determined simply at site by using an indicator solution which changes colour at a particular pH. Carbonation of concrete can take place quite quickly and it is therefore necessary to carry out the test on a freshly fractured surface. The most commonly used indicator is phenolphthalein which is colourless at a pH of less than 10 (carbonated concrete) and purple at higher values. A description of the procedure is given in a Building Research Establishment information paper.[4.23] The suggested indicator solution consists of 1 g of phenolphthalein dissolved in 50 ml of ethyl alcohol and then diluted to 100 ml with water. This is applied using a spray bottle to a freshly broken face which is approximately perpendicular to the surface of the member. The depth to the purple coloration is measured. If the carbonated layer is of fairly uniform thickness, the average depth is recorded. Otherwise the average depth and maximum depth, or the limits of the range of carbonation depths are recorded.

An alternative technique for determining carbonation depth, which is less destructive, is to drill holes in small increments using a masonry bit and spraying the hole with phenolphthalein after each increment. When the uncarbonated layer is reached, the bottom of the hole is coloured purple after spraying. One problem with this method is that the action of drilling may break up highly alkaline unhydrated cement grains within the carbonated region. This can cause a purple discoloration on the sides of the hole giving a false apparently low reading of carbonation depth. This difficulty can be minimized by carefully cleaning out the hole each time before spraying and also by carrying out calibration checks on some freshly fractured surfaces. The test sites should be chosen carefully, bearing in mind that carbonation depths may be increased at corners and at cracks as discussed in Chapter 2.

Radiography

Radiography employing X-rays or gamma rays is sometimes used to investigate defects such as honeycombing or poor compaction inside concrete members. The technique may also be used to determine the location of reinforcement in congested sections or other sections, where the use of the covermeter is not possible or is inappropriate. British Standard 1881: Part 205[4.24] gives recommendations for the use of the technique.

Gamma ray sources are usually employed for concrete sections up to 500 mm thick. High energy X-rays are appropriate for greater thicknesses. The source, consisting of Cobalt 60 or Iridium 192 for gamma rays or a linear accelerator for X-rays, is placed on one side of the member under investigation. A film, sensitive to X-rays, is placed in a light-tight cassette in contact with the other side of the member. Intensifying screens of lead foil mounted on card may be placed on either side of the films. The foil emits electrons when it is subjected to irradiation and this causes an intensification of the image, thus reducing the exposure time. The British standard suggests thicknesses of both front and back screens for X-ray and gamma ray sources. Source to film distances are also given for various source diameters to minimize lack of sharpness due to the configuration.

Strain

Strain in concrete is measured during load tests and also sometimes during service. When cracks develop in structures their movements can be monitored by a variety of means. The simplest method of assessing whether a crack is moving is to bond a glass strip (tell-tale) across the crack. The glass strip fractures if subsequent movement takes place. It is also possible to assess magnitude and direction of movement from the relative displacement of the two portions of the strip. This simple concept has been developed by Avonguard from whom a more sophisticated device is available. The device consists of two plastic strips which are attached one on each side of the crack so that they are able to slide over one another. One strip is marked with a grid and the other with a cross-hair. The position of the cross-hair on the grid is recorded and this allows the movement of the crack in two directions to be calculated.

Strains can be measured using mechanical, electrical resistance or vibrating wire gauges. Recommendations are given in BS 1881: Part 206.[4.25]

An instrument known as the demec (demountable mechanical) strain gauge is available[4.26] which consists of a bar with a fixed pin at one end and a pin on a pivot at the other, as shown in Fig. 4.16. Movements of the pin are magnified by a spring lever system and measured using a dial gauge. Circular metal studs with a central drilled hole are fixed on the concrete surface using an epoxy or other suitable adhesive. In exposed locations, the studs can be recessed into a small hole. The studs

Fig. 4.16
Demountable
mechanical strain
gauge

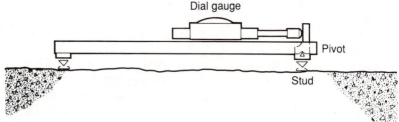

are set to the correct gauge length using a standard bar with two rigid pins similar to those on the gauge itself. Each time a reading, or set of readings, is taken they are compared with a reading of two studs on an invar bar. In this way temperature compensation is achieved. The method is relatively cheap and simple, and is extremely useful when only a few readings are to be taken on each occasion. When a larger number of readings has to be obtained or where access is difficult, a technique with remote read-out is more appropriate.

Electrical gauges of wire or foil use the principle that strain causes a change in resistance. They consist of a wire in a flat zigzag formation, or a foil etch cut to a similar shape bonded between two sheets of thin plastic. In the case of wire gauges there are two protruding wires for connection. Foil gauges have enlarged areas at the ends of the grid to which connecting leads can be soldered. The surface to which the gauge is to be applied is cleaned carefully so that it is free from dirt, grease, moisture, laitence, dust and other loose material. A special adhesive is used to bond the gauge to the surface, and care is taken to achieve good contact and to remove entrapped air. Gauges can be bonded to reinforcement as well as concrete.

The change in resistance caused by change in strain is measured using a Wheatstone bridge. Changes in temperature also cause change in resistance, and to take account of this a dummy gauge not subjected to strain is placed in a second arm of the bridge.

Vibrating wire gauges use the principle that the fundamental resonant frequency of a stretched wire depends on the tension, i.e. like the string of a musical instrument, the higher the tension the higher the note or frequency. In a vibrating wire or acoustic gauge the wire is contained in a rigid tube and plucked by a pulse through an electromagnet mounted near its centre. This same magnet is then used to detect the vibration, and its frequency is measured using an electronic instrument. The construction of a typical vibrating wire gauge is shown in Fig. 4.17.

Gauges are mounted on concrete using brackets firmly fixed to the surface. They can also be clamped directly onto reinforcing bars. It is possible to tune the gauges before taking the first reading so that gauges

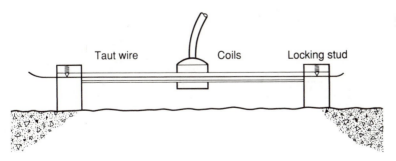

Fig. 4.17 Vibrating wire strain gauge

Taut wire   Coils   Locking stud

which are expected to experience compression start with a high frequency, and gauges that are expected to experience tension start off with a low frequency. Changes in temperature also affect the tension in the wire and it is usually necessary to check the change in frequency of dummy unstrained gauges so that corrections can be made.

Permeability

Several types of apparatus have been developed for site use in measuring properties related to permeability. The initial surface absorption test (ISAT)[4.27] uses a cap of surface area 5000 mm^2 sealed onto the surface of the concrete under test using modelling clay. When the test is undertaken on a vertical surface, a means of keeping the cap firmly in contact with the surface has to be provided. Water is introduced into the cap to give a pressure head of 200 mm using a filter funnel. A second port in the cap leads to a horizontal capillary tube. The rate at which water is absorbed into the concrete surface is determined by closing the connection to the reservoir and measuring the movement of water surface in the capillary tube during a fixed time period. On site the test is usually carried out 10 minutes after filling the cap with water. A diagram of the apparatus is given in Fig. 4.18.

A second test using either water or air has been developed by Figg[4.28] and modified by Cather et al.[4.29] For the original test a 5.5 mm diameter hole 30 mm deep is drilled into the concrete surface and carefully cleaned out. The hole is plugged near the surface and sealed with silicone rubber. For the water test, a double hypodermic needle is inserted into the hole through the seal and water is injected through the inner needle using a syringe. The water is forced back through the outer needle and into

Fig. 4.18 Initial surface absorption test

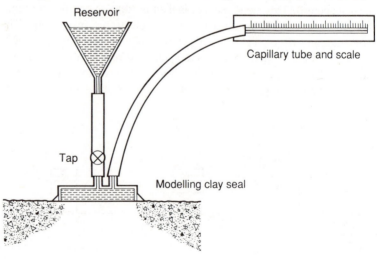

Reservoir

Capillary tube and scale

Tap

Modelling clay seal

a horizontal capillary tube positioned 100 mm above the base of the hole as shown in Fig. 4.19. The test is carried out by closing the valve to the syringe and determining the time for the meniscus to travel 50 mm along the tube. The air test is carried out in a similar way except that a vacuum pump is used to reduce the pressure in the hole to 15 kPa and a manometer is fitted in place of the capillary tube. The valve to the vacuum pump is closed and the time for the pressure to rise by 5 kPa is recorded.

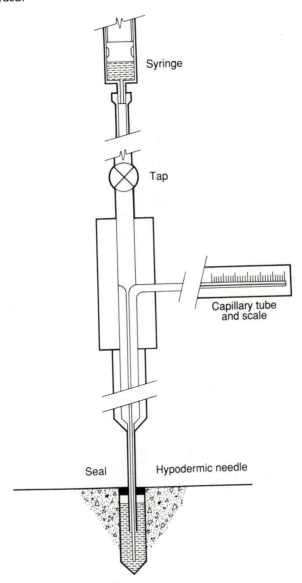

Fig. 4.19 Figg apparatus

Syringe

Tap

Capillary tube
and scale

Seal   Hypodermic needle

The Arup modification is aimed at producing a portable apparatus for use on site. It uses air only and a hole of diameter 10 mm and depth 40 mm. Pressure is measured using a digital electronic manometer and a standard filter is employed to protect the manometer from dust. The filter also serves the purpose of maintaining the total volume of air in the system to a figure similar to that in the original apparatus. A hand pump is used to reduce the pressure in the system and for this reason the test pressures have been changed. The time for a rise from 45 kPa to 50 kPa is recorded.

A third portable instrument called the Clam[4.30] uses a circular steel ring bonded to the concrete surface. The main body of the instrument is attached to the ring-using bolts. This creates a chamber in which water at constant pressure is applied to the surface (see Fig. 4.20). The pressure

Fig. 4.20 Clam apparatus

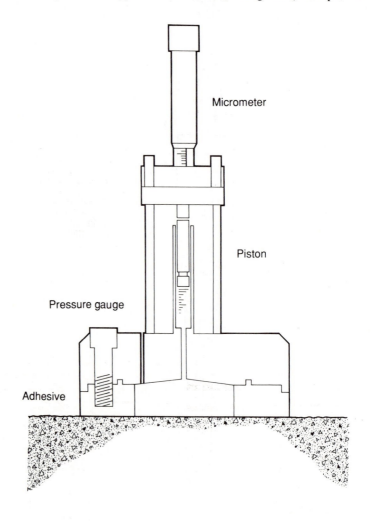

is maintained by means of a piston controlled by a micrometer screw gauge. Readings on the gauge are noted at the beginning and end of a test, which allows the volume of water which has flowed into the concrete to be calculated.

Laboratory Test Methods

Several laboratory test methods are available to assist in the assessment process. Suitable tests are described in Part 124 of BS 1881[4.31] whilst ASTM C 856[4.32] gives guidance on petrographic examination of samples of hardened concrete.

The samples for laboratory testing may be either cores, lump samples or drillings.

Cement Type

Determination of the type of cement used in construction is sometimes advantageous in assessing causes of deterioration and likely future performance. British Standard 1881: Part 124 describes chemical and microscopic methods.

The chemical method relies on the different compositions of varieties of cements. For example, sulphate-resisting cement has a much higher ratio of Fe_2O_3 to Al_2O_3 than ordinary Portland cement. The British standard gives typical analyses for four commonly available cements including Portland–blast-furnace and Portland–PFA. However, because there can be large variations in the chemical analysis of both blast-furnace slag and PFA and the proportions of these materials combined with cement can vary, there can be large deviations from the standard analysis for particular cements of these types.

The method relies on obtaining a sample which consists mainly of hydrated cement matrix. This is achieved by crushing the sample from site in a compression-testing machine and using a $75\mu m$ sieve to separate the finest material. The fine fraction is analysed for loss on ignition, insoluble residue and a range of the oxides normally found in cement — calcium oxide, soluble silica, alumina, ferrous oxide, magnesium oxide, etc. The loss on ignition is assumed to represent the combined water in the hydration products, and carbon dioxide if the sample is carbonated. The insoluble residue is taken to represent the aggregate present in the sample. The analyses of the proportions of the oxides in the sample are corrected to take account of these factors comparing them with the typical analyses given in the standard.

The microscopic method relies on examination of relics of unhydrated cement grains in a polished sample, and etching the specimen to reveal the phase composition. The test is usually carried out to determine whether sulphate-resisting or ordinary Portland cement was used in the production of the concrete.

The samples for examination are produced by breaking up the sample from site and selecting at least ten pieces consisting of a cement-rich matrix with a size of approximately 5 mm. These small pieces are dried at low temperature and then set in polyester or epoxy resin in a cylindrical mould. The casting is sectioned with a diamond saw using a non-aqueous lubricant so that cut faces of the concrete pieces are revealed. The cut surface is ground and polished using silicon carbide and diamond powders, again with non-aqueous lubricants. At this stage the polished surface is examined under a microscope to determine whether sufficient numbers of unhydrated cement grains of adequate size are present. The British standard recommends examination of at least 10 grains of size greater than $40\,\mu$m or at least 20 grains of size greater than $20\,\mu$n.

The surface of the sample is then etched with either potassium hydroxide solution or hydrofluoric acid vapour. Etching reveals mineral phases and their relative abundance can be used to assess the cement type. For example, etching with potassium hydroxide solution causes the tricalcium aluminate phase to appear bluish grey in colour, while the silicates remain grey and the ferrite appears brilliant white. The volume ratio of ferrite to tricalcium aluminate can be used as an indication of whether an ordinary Portland cement or a sulphate-resisting cement is present. Clinker grains from a particular cement are not consistent and occasional anomalies such as grains with high tricalcium aluminate content in a sulphate-resisting cement may be found. However, usually a minimum of 80 per cent of unhydrated grains of reasonable size should indicate either an ordinary Portland or sulphate-resisting cement.

British Standard 1881: Part 6,[4.33] which is now superseded, gave details of an etching procedure for identification of high alumina cement. According to this the polished surface is etched with boiling potassium hydroxide solution. The presence of high alumina cement is then indicated by blue or brown irregular grains which exhibit twinning.

Cement Content

The method for determination of cement content also relies on chemical analysis and a knowledge of the composition of cement. A sample of the concrete is crushed and ground in stages until a representative sample passing a $150\,\mu$m sieve is obtained. The soluble silica and calcium oxide are extracted using dilute hydrochloric acid, and the amounts of these compounds present are determined by analytical chemical means. The cement content of the concrete samples can be calculated from both the silica content and the calcium oxide content.

In most cases when an old building is being investigated, the actual composition of the cement will not be known and it will be necessary to use assumed values for the soluble silica and calcium oxide contents. The assumed values given in Part 124 of BS 1881 are 64.5 per cent for calcium oxide and 20.2 per cent for silica. If the cement contents calculated from the two compounds are within 1 per cent, the mean value is taken. When the calculated cement contents differ by more than 1 per cent the reason for the discrepancy is investigated. Calcium oxide or soluble silica in the aggregates can cause discrepancies of this sort. If samples of the aggregate are available, it is possible to determine their calcium oxide and soluble silica contents, and correct for them. However, this will rarely be possible in the case of investigation of concrete in service. In this case the lower of the two calculated values is usually taken.

The above procedure is appropriate for concrete containing ordinary Portland and sulphate-resisting cements. Pulverized fuel ash and ground granulated blast-furnace slag are used in some cements, and they also contain soluble silica and calcium oxide. If they are present in the concrete sample it may not be possible to obtain an accurate determination of the total cementitious material content by chemical methods. One approach for concretes containing blast-furnace slag cement is to determine the sulphide content in addition to the calcium oxide and soluble silica. If the composition of the blast-furnace slag is known, the results of the analysis can be corrected but the accuracy of the results will remain in doubt.

Concrete Society Technical Report No 32[4.34] suggests that for mixes containing ordinary Portland or sulphate-resistant cement, the accuracy of the measured cement content at 95 per cent confidence limits is $\pm 25 \, \text{kg m}^{-3}$ if four independent samples are analysed. The above figures do not include errors arising from the analysis itself and are based on a standard deviation of $25 \, \text{kg m}^{-3}$. An appendix to the report gives details of a precision experiment in which 18 laboratories analysed concretes with cement contents in the range $240-425 \, \text{kg m}^{-3}$, and containing Thames valley flint or Ballidan limestone aggregate. Brief details of the mixes and the range of results obtained are shown in Table 4.2. The results shown in this table are those obtained without analysing

Table 4.2 Range of results obtained in a precision experiment on cement content determination (based on Concrete Society report[4.34])

Concrete sample	Aggregate	Cement : Fine : Coarse	Water/cement ratio	Cement (kg m^{-3})	Range CaO analysis	Range SiO_2 analysis
1A	Flint	1 : 3.5 : 4.7	0.73	240	268–350	268–438
1B	Flint	1 : 3.2 : 4.3	0.67	240	181–349	275–545
2A	Flint	1 : 1.5 : 2.7	0.42	425	403–580	404–617
2B	Flint	1 : 1.4 : 2.6	0.40	425	345–575	449–640
3A	Limestone	1 : 2.1 : 3.5	0.59	345		301–549
3B	Limestone	1 : 2.0 : 3.3	0.55	345		327–582

samples of the aggregate. It can be seen that a wide range of values of cement content were reported for each of the concretes analysed.

Original Water Content

Only part of the water present in a fresh concrete mix takes part in the hydration reaction. The remainder fills the capillary pores which are formed as the concrete sets. An estimate of the original water content of a concrete sample can be obtained by determining the capillary porosity and the combined water, and adding the results.[4.31] Capillary porosity is determined using a sample of concrete with two parallel saw-cut faces. The sample is approximately 20 mm thick and has a single face area of not less than 10 000 mm. The sample is first dried at 105 °C and then immersed in trichloroethane in a vacuum desiccator. When the vacuum is applied, the air leaves the pores and is replaced by trichloroethane. The increase in weight allows the volume of the pores to be calculated. Combined water is determined by heating a powdered sample of concrete to 1000 °C in a combustion tube in a stream of dry nitrogen. After leaving the combustion tube, the stream of nitrogen is passed through two absorption tubes packed with dried magnesium perchlorate. The increase in mass of the absorption tubes is taken as the mass of combined water in the sample.

Chloride Content

Samples for chloride content determination are usually in the form of cores, lumps or drill dust. In the former two cases, it is necessary to break down and grind the sample before carrying out chemical analysis. This is done in a similar manner to that described for cement content determination above. An alternative is to prepare a sample from the matrix alone, discarding the coarse aggregate. If this is the case, a cement content determination is undertaken on a portion of the sample also.

American and British practice differs in the way that chloride is extracted from the sample. In the United States the practice adopted by the Federal Highway Administration is to boil the sample in water for 5 minutes and then to let it stand for 24 hours before chemical analysis. British practice[4.31] is to extract using nitric acid. Having dissolved the chloride from the sample, the concentration is determined by titration.

Approximate methods of determining chloride contents at site have been suggested in a Building Research Establishment information sheet.[4.35] The fluid obtained by extraction is tested using titrator strips (Quantabs) or a chloride test kit (Hach). Titrator strips consist of a capillary tube filled with a compound which reacts with chloride in solution and changes colour. When the titrator strip is placed in a chloride solution, the fluid is drawn up the tube. The length of colour change in the tube is a measure of the concentration of chloride in the solution. Each batch of titrator strips has its own conversion chart giving the relationship between length of colour change in the tube and chloride concentration. When using a chloride test kit, a known volume of the solution is taken. A test fluid is added to the sample drop by drop from a dropper bottle. The number of drops to cause a colour change is related to the chloride content of the sample.

Sulphate Content

Sulphate content samples are prepared in the same way as samples for chloride determination. In this case the extraction is carried out using concentrated hydrochloric acid.[4.31] The solution is neutralized using dilute ammonium hydroxide and then barium chloride is added to produce a precipitate of barium sulphate. The weight of barium sulphate produced permits the sulphate content of the concrete sample to be calculated.

Petrographic Examination

Petrography is the section of geology dealing with description and classification of rocks, but petrographic techniques can also be used to inspect samples of concrete in the laboratory. A standard practice is described in ASTM C 856[4.32] and a description of microscopic methods is given in Concrete Society Technical Report No 32[4.34]. Areas in which microscopic methods have been found to be useful include:

1. Identification of mix ingredients including coarse and fine aggregate type, cement type, pulverized fuel ash and ground granulated blast-furnace slag.
2. Assessment of mix proportions and air void content.
3. Investigation of various durability problems including carbonation, alkali—aggregate reaction, fire damage and chemical attack.

The actual petrographic procedures to be adopted in examination of a particular concrete sample will develop as the examination proceeds and a flexible rather than a rigid standardized approach is appropriate. ASTM C 856 therefore gives guidance and recommended procedures. Three different types of microscopic examination may be employed.

The initial stage usually involves examination of specimens as-received using the naked eye and a stereoscopic microscope, typically at a magnification up to ×100. The examination should yield useful information on aggregate type and the presence of any reaction rims. It will also permit crack and void system to be examined, particularly for the presence of reaction products. These could include the gel that is a product of alkali—silica reaction, ettringite which results from sulphate attack and calcite, or calcium hydroxide which may be evidence of leaching. Stereoscopic microscopic examination yields useful information in its own right, but another of its functions is to provide information to enable the location of thin section and polished section samples to be chosen.

The polarizing petrological microscope is a very powerful analytical tool which can be used to positively identify mineral constituents of aggregates, cement paste and the products of deterioration from their optical properties. Magnification employed is usually in the range $\times 5$ to $\times 50$, though higher magnification is possible. Samples are examined in thin section, typically $20-40\ \mu m$ thick. These thin sections are prepared by first cutting a slice of approximately 2 mm thick using a diamond saw. In some cases where the concrete is weak or friable, it may be necessary to vacuum impregnate the sample with resin prior to slicing. In all of the processes of lapping and polishing involved in producing the thin section, the use of water or production of heat has to be avoided as these may modify the mineral structures which are present. One face of the wafer is polished smooth and the specimen is mounted on a glass plate. The section is ground down using progressively finer abrasives until the desired thickness is obtained.

The examination is then carried out using polarized light transmitted through the specimen. The birefringent properties of various minerals allows them to be identified. Thin sections also permit the presence of pulverized fuel ash or ground granulated blast-furnace slag to be identified. Particles of unhydrated pulverized fuel ash present in concrete have a spherical shape and a glassy appearance, remnant particular ground granulated blast-furnace slag is glassy, angular and shard-like. Natural pozzolanas are used in the production of cement in many countries and their presence in concrete can also be detected by examination of thin sections.

The metallurgical (metallographic in the United States) microscope is used to examine polished specimens under reflected light to a magnification of $\times 500$. A petrological microscope can be used if it is fitted with an incident light attachment. The principal use of the metallurgical microscope in concrete examination is in the identification of cement type as discussed earlier in this chapter.

Petrographic techniques can be used in a quantitative way to assess the original mix proportions. Two methods using either point counting or linear traversing of polished sections are commonly employed. The former is easier and more rapid than the latter. The basic principles of both methods are described in ASTM C 457[4.36] in relation to the determination of air void content and bubble size and spacing. In each of the methods the polished specimen is placed on a hand or motor-driven stage which can be moved in two directions at right angles. This permits the specimen to be traversed along regularly-spaced parallel lines.

In the point count method, the movement is stopped at regular points along each line and a tally counter is used to record which of the components of concrete, e.g. coarse aggregate, fine aggregate, cement paste or void, lies at the position of the cross-hairs. For the linear traverse method the length traversed across each component is measured and integrated as the cross-hairs move along regularly-spaced parallel lines. Using either of these methods, the volumetric percentages of each of the components can be determined. With the measurement of some additional parameters the cement content, water/cement ratio and other details of the original mix can be assessed.

Alkali Reactivity

The diagnosis of alkali reactivity involves a number of testing and inspection techniques, including identification of crack patterns and petrography as described in earlier sections. The Institution of Structural Engineers, in a report[4.37] on the structural effects of alkali–silica reactivity, have suggested a programme of site measurements and laboratory testing in order to arrive at an 'expansion index' which is used when making decisions on the management of affected structures. The current expansion is assessed by summing the widths of cracks measured on the structure and the potential for additional expansion is determined by exposing cores from the structure at 100 per cent relative humidity.

Current Expansion

The tentative recommendation of the report is that the existing expansion in the structure should be assessed by drawing a straight line on the surface, measuring the widths of all cracks crossed by the line and adding these together. It is suggested that at least five lines, not less than 1 m long and 250 mm apart, are drawn on the surface in the direction at right angles to the main crack direction. The lines are drawn on the face of the member which is most severely affected. The result is expressed as a strain by adding together the crack widths and dividing by the length of the line.

Potential Additional Expansion

Expansion of cores, even from one pour of concrete, is extremely variable. To limit the effects of this variability, the report suggests that testing is carried out on at least four cores taken from different pours. Demec studs (see earlier section on strain measurement) are fixed to give gauge lengths at three equally-spaced positions on the circumference. The cores are stored at 38 °C and 100 per cent relative humidity for a period of at least 6 months, and the expansion measured at regular intervals. The water uptake is also determined by the increase in weight. For structural assessments, the conditions in place in the structure can be modelled more closely by carrying out the test at 5, 13 or 20 °C. It has been found with some aggregates that although the initial expansion is less rapid when the test is carried out at lower temperatures, the total expansion at greater ages is substantially higher. It is suggested that the test should be concluded after say 6 months and the potential for additional expansion assessed by extrapolating the curve to 2 years.

References

4.1 Yip W K, Tam C T 1988 Concrete strength evaluation through the use of small diameter cores. *Magazine of Concrete Research* **40 (143):** 99−105

4.2 British Standards Institution 1983 *Testing concrete. Method for Determination of the Compressive Strength of Concrete Cores* BS 1881: Part 120, The British Standards Institution, London

4.3 American Society for Testing and Materials 1984 *Standard Method of Obtaining and Testing Drilled Cores and Sawed Beams of Concrete* ASTM C 42, The American Society for Testing and Materials, Philadelphia

4.4 American Society for Testing and Materials 1985 *Method of Capping Cylindrical Test Specimens* ASTM C 617, The American Society for Testing and Materials, Philadelphia

4.5 ACI Committee 318, 1983 (revised 1986) *Building Code Requirements for Reinforced Concrete* ACI Standard 318-83, American Concrete Institute, Detroit

4.6 British Standards Institution 1986 *Testing Concrete. Recommendations for Surface Hardness Testing by Rebound Hammer* BS 1881: Part 202, The British Standards Institution, London

4.7 American Society for Testing and Materials 1985 *Test Method for Rebound Number of Hardened Concrete* ASTM C 805, The American Society for Testing and Materials, Philadelphia

4.8 ACI Committee 228 1988 In-place methods for determination of strength of concrete. *ACI Materials Journal* **Sept/Oct:** 446−71

4.9 British Standards Institution 1986 *Testing Concrete. Recommendations for Measurement of Velocity of Ultrasonic Pulses in Concrete* BS 1881: Part 203, The British Standards Institution, London

4.10 American Society for Testing and Materials 1983 *Test Method for Pulse Velocity Through Concrete* ASTM C 597, The American Society for Testing and Materials, Philadelphia

4.11 American Society for Testing and Materials 1982 *Test Method for Penetration Resistance of Hardened Concrete* ASTM C 803, The American Society for Testing and Materials, Philadelphia

4.12 British Standards Institution *Testing Concrete. Recommendations for the Assessment of Concrete Strength by Near-to-Surface Tests* BS 1881: Part 207, The British Standards Institution, London (in preparation)

4.13 British Standards Institution 1986 *Testing Concrete. Guide to the Use of Non-destructive Methods of Test for Hardened Concrete* BS 1881: Part 201, The British Standards Institution, London

4.14 Chabowski A J, Bryden-Smith D W 1980 *Internal Fracture Testing of In Situ Concrete: A Method of Assessing Compressive Strength* Information Paper 22/80, Building Research Establishment, Garston, United Kingdom

4.15 American Society for Testing and Materials 1987 *Standard Test Method for Pull Out Strength of Hardened Concrete* ASTM C 900, The American Society for Testing and Materials, Philadelphia

4.16 Long A E, Murray A McC 1984 The pull-off partially destructive test for concrete. In *In Situ/Non-destructive Testing of Concrete* ACI SP-82, The American Concrete Institute, Detroit pp 327–50

4.17 Carlsson M, Eeg I R, Jahren P 1984 Field experience in the use of the break-off tester. In *In Situ/Non-destructive Testing of Concrete* ACI SP-82, The American Concrete Institute, Detroit pp 277–92

4.18 Dahl-Jorgansen E, Johansen R 1984 General and specialised use of break-off concrete strength testing method. In *In Situ/Non-destructive Testing of Concrete* ACI SP-82, The American Concrete Institute, Detroit pp 293–308

4.19 British Standards Institution 1988 *Testing Concrete. Recommendations on the Use of Electromagnetic Covermeters* BS 1881: Part 204, The British Standards Institution, London

4.20 American Society for Testing and Materials 1980 *Standard Test Method for Half-cell Potentials of Reinforcing Steel in Concrete* ASTM C 876, The American Society for Testing and Materials, Philadelphia

4.21 Institute of Corrosion 1990 *Outline Specification For Equipotential Mapping Surveys* Draft document for comment, Institute of Corrosion, Leighton Buzzard

4.22 Institute of Corrosion 1990 *A Guide to Potential (Half-cell) Mapping* Draft document for comment, Institute of Corrosion, Leighton Buzzard

4.23 Roberts M H 1981 *Carbonation of Concrete Made with Dense Natural Aggregates* Information Paper 6/81, The Building Research Establishment, Garston, United Kingdom

4.24 British Standards Institution 1986 *Testing Concrete. Recommendations for Radiography of Concrete* BS 1881: Part 205, The British Standards Institution, London

4.25 British Standards Institution 1986 *Testing Concrete. Recommendations for Determination of Strain in Concrete* BS 1881: Part 206, The British Standards Institution, London

4.26 Morice P B, Base G D 1953 The design and use of a mechanical strain gauge for concrete structures. *The Magazine of Concrete Research* **5 (13)**

4.27 British Standards Institution 1970. *Methods of Testing Concrete. Methods of Testing Hardened Concrete for Other Than Strength* BS 1881: Part 5, The British Standards Institution, London

4.28 Figg J W 1973 Methods of measuring the air and water permeability of concrete. *Magazine of Concrete Research* **25 (85):** 213–19

4.29 Cather R, Figg J W, Marsden A F, O'Brien T P 1984 Improvements to the Figg method for determining the air permeability of concrete. *Magazine of Concrete Research* **36 (129):** 241–5

4.30 Basheer P A M, Montgomery F R, Long A E 1990 *In-situ*/non-destructive assessment of near surface permeability of concrete with 'Universal Clam' *Spring Convention*, American Concrete Institute, Toronto

4.31 British Standards Institution 1988 *Testing Concrete. Methods for Analysis of Hardened Concrete* BS 1881: Part 124, The British Standards Institution, London

4.32 American Society for Testing and Materials 1984 (reapproved 1988) *Standard Practice for Petrographic Examination of Hardened Concrete* ASTM C 856, The American Society for Testing and Materials, Philadelphia

4.33 British Standards Institution 1971 *Methods of Testing Concrete. Analysis of Hardened Concrete* BS 1881: Part 6, The British Standards Institution, London

4.34 Concrete Society 1989 *Analysis of Hardened Concrete* Technical Report No 32, The Concrete Society, Wexham, United Kingdom

4.35 Roberts M H 1986 *Determination of the Chloride and Cement Contents of Hardened Concrete* Information Paper IP 21/86, Building Research Establishment, Garston, United Kingdom

4.36 American Society for Testing and Materials 1982 *Standard Practice of Microscopical Determination of Air Void Content and Parameters of the Air Void System in Hardened Concrete* ASTM C 457, The American Society for Testing and Materials, Philadelphia

4.37 Doran D K (Chairman) 1988 *Structural Effects of Alkali–silica Reaction – Interim Technical Guidance on Appraisal of Existing Structures* Report of an *ad hoc* committee of the Institution of Structural Engineers, London

5   Interpretation of Results

The methods used to interpret the results of an inspection and survey depend upon the overall purpose of the investigation and the type, quality and amount of data obtained. Budget considerations may have constrained the amount of information collected and one of the decisions that has to be made is whether sufficient data has been obtained. The main results of the interpretation will usually be an assessment of the current condition of the building or structure, indications of its likely future performance and recommendations for immediate or future actions.

Assessments of current condition and considerations of future performance need to be carried out on the basis of the client's needs. As an example, results of load tests should be assessed taking into account the loading intensity appropriate to current or intended use and likewise assessments of future performance in durability terms should take into account differences in exposure conditions likely to be caused by any proposed change of use. Examples of recommendations for immediate or future actions include requirements for propping, the need to undertake further investigations or monitoring and recommendations to carry out repair or protection works. The assessment should also include observations on the causes and mechanisms which underlie any shortfalls in the current condition and comments on possible effect on future behaviour.

The interpretation of results of investigations is a complex matter and, in many cases, there appears to be little codified guidance on the reasoning processes to be employed. There are some notable exceptions to this generalization, for instance, the report on the appraisal of existing structures[5.1] by the Institution of Structural Engineers and also the Department of the Environment report[5.2] on structures containing high alumina cement concrete. Individual limiting or critical values have been specified for many of the properties measured in investigations but these values more usually relate to new construction and are not always appropriate to the case of assessing a structure in service. As an example, the limits given for chloride concentrations in fresh concrete may not be appropriate to the case where chlorides have entered concrete from an external source. This is because, in the latter case, there is less likelihood of the chlorides taking part in chemical reaction and being bound into the products of hydration; the chlorides are, therefore, more mobile and more likely to initiate corrosion of reinforcement.

Another difficulty is that, although limiting values for individual properties may be available, little is known in most cases about how two or more mechanisms can interact to affect the future performance of the structure. The properties may also be subject to considerable localized variations. Because of these difficulties, a two step approach to interpretation is often adopted in which the first step is to compare the results of individual tests against known critical values. This helps to identify locations where there are possible problems. The second step is to take an overview of the results of all of the tests in both the localized and structure-wide contexts.

This two-step approach leads to a generalized assessment of the condition of the structure under consideration, but extrapolating to give predictions of future behaviour is even more difficult. Most deterioration processes are time-related but the relationship may not be clearly understood even under steady-state conditions. In many practical situations the relationship between condition and time is affected by changes in other environmental agencies.

As a result, it is only possible in most cases to give a general indication of the present condition of the structure and very generalized predictions of the future performance of the structure within a fairly wide time-scale. This being the case and provided that safety of the structure itself and those who use it is not compromised, there may not be a single 'correct' solution for future action. It is, therefore, desirable to involve the structure's owner at an early stage, as the decision as to whether to demolish and reconstruct or to repair now or in the future may depend on commercial rather than technical judgements. As an example, industrial processes often have short useful lives and the associated structures may only be required to be sustained for a few years. The owner of the structure should therefore be consulted at various points during investigation and not merely at the end, so that there is an opportunity of curtailing the investigation if it is clear that the structure is unsatisfactory. On the other hand, the owner may choose to enlarge the scope if, by so doing, there is a greater chance of being able to clarify the extent of necessary repairs, or of being able to demonstrate the likelihood of adequate future performance.

In this chapter the means of interpreting and assessing the results from the tests described in the previous chapter are discussed. The assessment will be described on the basis of the results from an individual test method and also on the basis of combinations of test methods where this is possible.

Strength Tests

As discussed in Chapter 4, the strength of concrete in structures can be measured directly by cutting and crushing cores or indirectly by measuring a property for which a relationship with strength can be established. In most cases, where indirect methods are used, the relationship should ideally be established for the particular concrete under consideration as the relationship may change for different aggregates or constituents of the mix. When the results of indirect tests are being interpreted the accuracy of the relationship should be kept in mind. It may only be possible to place an individual result in a range identified by an upper and lower bound at a certain confidence level as shown in Fig. 5.1.

When strength values have been determined they are usually used to calculate a characteristic strength for the concrete under consideration. Characteristic strength is defined as that value of strength that would be expected to be exceeded with a certain percentage probability and is calculated statistically from the mean and the standard deviation assuming a normal distribution.

$$f_k = f_m - K(SD) \qquad\qquad [5.1]$$

where f_k = characteristic strength;
f_m = mean strength;
SD = standard deviation; and
K is a factor dependent upon the percentage probability (1.648 for 95 per cent probability).

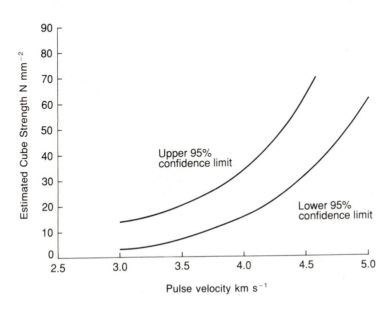

Fig. 5.1 Upper and lower confidence limit curves for experimentally derived relationship between ultrasonic pulse velocity and cube strength

However, before applying the formula, it is worth considering whether there are any obvious patterns to the variations in the strength of the concrete represented by the samples. For example, is it possible that a different grade of concrete may have been specified in different elements, or that a different final strength has been achieved in some elements because of differences in the difficulty of achieving adequate compaction or in providing efficient curing. If there are sufficient results, it is, therefore, worthwhile calculating means and standard deviations for groups of like elements which make up a structure. If these prove to be similar the results can be grouped together and a single global characteristic strength calculated. If not, the groups of results from different elements should be considered individually and separate characteristic strengths calculated.

Distribution Histograms

It is also often helpful to draw strength distribution histograms showing the number of results in particular strength ranges as they may draw attention to anomalies in the data. They are also useful in indicating the standard of quality control that was achieved during construction. Ideally, the histogram should approximate to a normal distribution with a pronounced peak and narrow spread as shown in Fig. 5.2(a). However, in practice it may often have a flatter peak and wider spread indicating poorer quality control.

In some cases as stated above, there may be two obvious peaks in the data as shown in Fig. 5.2(b). In these circumstances it should be possible to divide the results into two separate families by examination of the core locations or the cores themselves. If this is the case a characteristic strength can be calculated for each of the families.

Log—normal Distribution

Strength results and the results of other physical measurements at site may have a pronounced positive skew with a tail of higher values as shown in Fig. 5.2(c). If characteristic values are calculated from these results using Eq. 5.1 an unrealistically low value will be obtained because of the high standard deviation. The results may be a better fit to a log—normal distribution, and if this is the case an alternative approach to the calculation of characteristic value may be appropriate.[5.3] If the individual values are $f_1, f_2, f_3, \ldots f_n$, the natural logarithms of these values $\ln(f_1)$, $\ln(f_2)$, $\ln(f_3)$, $\ldots$ $\ln(f_n)$ approximate to a normal distribution with mean $\ln(f_m)$ and standard deviation $SD_{\ln f}$. It follows from Eq. 5.1

$$\ln(f_k) = \ln(f_m) - K(SD)_{\ln f} \qquad [5.2]$$

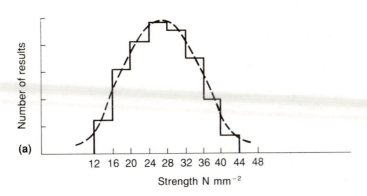

Fig. 5.2 Strength
distribution
histograms.
(a) distribution of
strength data and
approximation to
normal curve;
(b) distribution of
strength data for
two overlapping
populations;
(c) distribution of
strength data
showing positive
skew

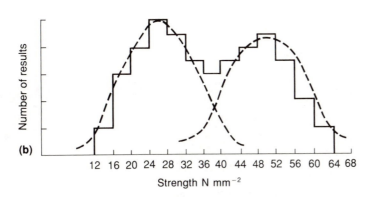

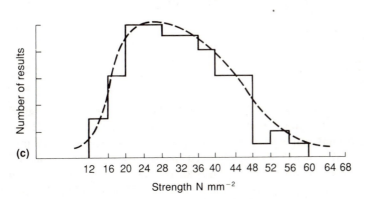

where $\ln(f_k)$ is the natural logarithm of the characteristic strength; and $SD_{\ln f}$ is the standard deviation of the log strength values

and therefore

$$f_k = \exp(\ln(f_m) - K(SD)_{\ln f}) \qquad\qquad [5.3]$$

Table 5.1 Strength results used in
Fig. 5.3 (Strength results: f_m =
28.5 N mm^{-2}, SD = 9.3 N mm^{-1},
f_k = 13.0 N mm^{-1}; natural
logarithm values: mean = 3.31,
SD = 0.29, f_k = 17.0 N mm^{-2})

Strength result f	$\ln(f)$
16.0	2.77
17.5	2.86
18.0	2.89
20.0	3.00
21.0	3.04
22.0	3.09
22.5	3.11
23.0	3.14
23.0	3.14
24.0	3.18
25.0	3.22
26.0	3.26
26.0	3.26
27.0	3.30
27.5	3.31
28.0	3.33
28.0	3.33
29.0	3.37
30.0	3.40
31.0	3.43
31.5	3.45
32.0	3.47
32.0	3.47
33.0	3.50
35.0	3.55
37.0	3.61
39.0	3.66
48.0	3.87
61.0	4.11

The procedure is illustrated by the data in Table 5.1. The figures in the first column are the strength results as received. When plotted in 5 N mm^{-2} strength bands, they give the distribution shown in Fig. 5.3(a), which is positively skewed and has two isolated high values. The second column of Table 5.1 lists the natural logarithms of the strengths. When these are plotted in increments of 0.2 the distribution shown in Fig. 5.3(b) results. This histogram is a better approximation to a normal distribution. If Eq. 5.1 is applied to the basic strength data it results in a characteristic strength of 13.0 N mm^{-2} which is 3.0 N mm^{-2} less than the lowest test result. Using the natural logarithms of the results and Eq. 5.3 leads to an enhanced value of 17.0 N mm^{-2}. Since only one of the 29 values (i.e. 3.4 per cent) is less than 17.0 N mm^{-2}, this would appear to be a more realistic figure for use in structural calculations.

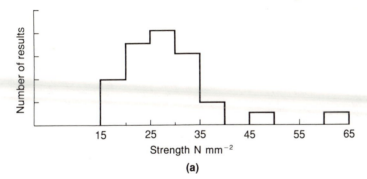

Fig. 5.3 Data from
Table 5.1 plotted
as a frequency
distribution.
(a) values as
recorded;
(b) natural
logarithms of
recorded values

(a)

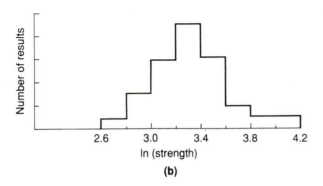

(b)

Probability Distribution

The Hong Kong Housing Authority[5.4] have used a different approach
to the analysis of a set of strength test results which do not follow a normal
distribution. The results are plotted on probability paper and a smooth
curve is drawn through the points. The data from Table 5.1 have been
used to produce Fig. 5.4, where the cumulative probability for a particular
strength is plotted as the percentage of results less than that value. The
value on the curve where it cuts the 5 per cent probability line can be
taken as the characteristic strength. On Fig. 5.4 the value produced is
approximately $18 \, \text{N mm}^{-2}$ which compares favourably with $17 \, \text{N mm}^{-2}$
resulting from calculation using Eq. 5.3.

Broken Cores

Much of the investigation work carried out by the Hong Kong Housing
Authority has been on concrete of low strength, and problems were
experienced with a significant number of cores which broke during
extraction and which could not be tested for strength. These problems
were overcome by using a procedure in which strengths were assigned
to broken cores. All broken and unbroken cores were carefully examined

Fig. 5.4 Data from
Table 5.1 plotted
on probability
paper

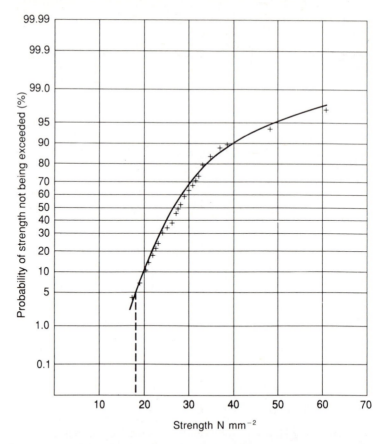

Fig. 5.4 Data from Table 5.1 plotted on probability paper

by two experienced engineers who ranked the cores for strength on a
scale of 1 to 10. Spearman's rank correlation method was used to compare
the rankings from the two engineers. Agreement between the two
engineers was found to be very significant. All cores with sufficient length
were crushed and the strengths were compared with the mean ranking
of the judges, by producing a scatter plot. This showed a reasonable
correlation between strength and ranking and permitted a strength to be
assigned to broken cores using a non-linear regression line. This would
seem to be a useful technique when a significant number of results are
being processed.

Bayes' Theorem

A further approach to assessment of characteristic strength from core
results is provided by Bayes' theorem.[5.5] Thomas Bayes was a minister
of religion and mathematician who lived in England during the eighteenth
century. His theorem can be written in the form:

$$P(S_i|D) = P(D|S_i) \times P(S_i)/\Sigma\{P(D|S_n) \times P(S_n)\} \qquad [5.4]$$

where $P(\;)$ is read as probability of the bracketed expression;

S_i is a state of an unknown quantity;

S_n is the set of all of the states of the unknown quantity;

D is the observed data;

| is read as given.

The expression appears to be highly complex, but in practice it can be used in quite a straightforward way, particularly with the help of a spreadsheet computer program. First a set of possible values for the characteristic strength of the concrete represented by the samples is selected. It is assumed that each characteristic strength has an equal probability of occurring. In order to apply the theorem, it is necessary to be able to calculate the probability of a particular value of strength being measured for each of the selected values of characteristic strength. This can be done if the standard deviation is known from past experience or from published data.

The procedure is best illustrated by example. If the particular value of characteristic strength being tested is $20\,N\,mm^{-2}$ and a value of standard deviation of $5.0\,N\,mm^{-2}$ is used, the concrete would have had a mean strength given by:

$$20 + 1.648 \times 5.0 = 28.24\ N\ mm^{-2}$$

The normal distribution of test results for a concrete with a mean strength of $28.24\,N\,mm^{-2}$ and standard deviation of $5.0\,N\,mm^{-2}$ is shown in Fig. 5.5. It can be seen that there is a high probability of a $30.0\,N\,mm^{-2}$ strength result belonging to this distribution as it lies close to the mean value. There is less probability of a $23.0\,N\,mm^{-2}$ result

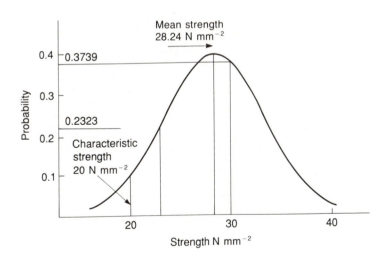

Fig. 5.5 Probability of obtaining particular strength results from a concrete with a characteristic strength of $20\ N\ mm^{-2}$

belonging to the distribution because it is further away from the mean value and the frequency ordinate is smaller. The probability that a particular result belongs to a particular distribution is proportional to the frequency ordinate at that value, and is given by:

$$f = 1/\{\sqrt{(2\pi)} \times e^a\} \qquad [5.5]$$

where u = (strength value − mean value)/standard deviation.
     e is the base of natural logarithms
     $a = u^2/2$

The above case where the result is 23.0 N mm^{-2}, the mean value is 28.24 N mm^{-2} and the standard deviation is 5.0 N mm^{-2} gives:

$$u = 1.04 \text{ and hence}$$
$$f = 0.2323$$

When applying Bayes' theorem, the probability of each of the cores belonging to a frequency distribution associated with each of ten chosen characteristic strengths from 16 to 25 N mm^{-2} is calculated as shown in Table 5.2. For the example, three cores have been chosen at random from the 29 results given in Table 5.1. The probabilities for each core are multiplied together to give an overall probability for each

Table 5.2 Calculation of probability of characteristic strength from strength results using Bayes' theorem[5.5]

Characteristic strength (N mm^{-2})	Standard deviation (N mm^{-2})	Mean strength (N mm^{-2})	30		23		29		L	P
			u	f	u	f	u	f		
16	5	24.2	1.16	0.2036	0.24	0.3876	0.96	0.2516	0.0199	0.087
17	5	25.2	0.96	0.2516	0.44	0.3621	0.76	0.2989	0.0272	0.119
18	5	26.2	0.76	0.2989	0.64	0.3251	0.56	0.3410	0.0331	0.145
19	5	27.2	0.56	0.3410	0.84	0.2803	0.36	0.3739	0.0357	0.156
20	5	28.2	0.36	0.3739	1.04	0.2323	0.16	0.3939	0.0342	0.149
21	5	29.2	0.16	0.3939	1.24	0.1849	0.04	0.3986	0.0290	0.127
22	5	30.2	0.04	0.3986	1.44	0.1414	0.24	0.3876	0.0218	0.095
23	5	31.2	0.24	0.3876	1.64	0.1039	0.44	0.3621	0.0146	0.064
24	5	32.2	0.44	0.3621	1.84	0.0734	0.64	0.3251	0.0086	0.037
25	5	33.2	0.64	0.3251	2.04	0.0498	0.84	0.2803	0.0045	0.020

Notes: 1. Mean strength = characteristic strength + 1.648 × (standard deviation)
     2. u = (strength result − mean strength)/standard deviation
     3. $f = 1/\{\sqrt{(2\pi)} \times e\dagger(u^2/2)\}$
     4. L is the likelihood of a particular characteristic strength given by the product of the three fs. If it had been possible to make an initial assessment of probability these values would be included in the product. In this case, the initial probabilities were considered to be equal and therefore had a value of 0.1. They have been omitted for simplicity.
     5. P is the probability of a particular characteristic strength given by dividing the Ls by the sum of the Ls.

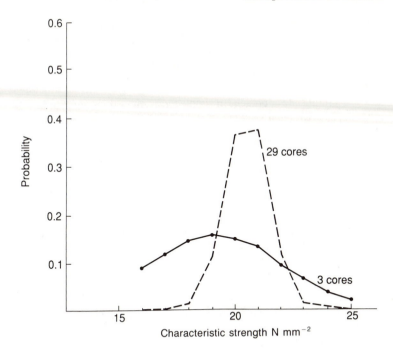

Fig. 5.6 Probability of characteristic strength calculated using Bayes' theorem and the data in Table 5.1

characteristic strength. As can be seen from Table 5.2 and Fig. 5.6, the most probable values for characteristic strength lie in the range $17-21 \, \text{N mm}^{-2}$ but no single value in this range has a particularly high probability.

If the calculation is carried out using all 29 results, the calculated probabilities are as shown also on Fig. 5.6. The 20 and $21 \, \text{N mm}^{-2}$ characteristic strength values have high probabilities (0.365 and 0.375), whereas the next highest value is 0.12. The indicated characteristic strength is therefore somewhat higher than the values calculated previously by using a log−normal distribution and by plotting the probability curve. This is understandable in view of the skew distribution of the 29 results and the assumptions made in applying Bayes' theorem. Bayes' theorem is probably of more use for assessing characteristic strengths from a small number of results when something is known about the control exercised during construction.

Structural Assessment

The calculations carried out for assessment of an existing structure are akin to those used in the design process. Firstly, the structure, or a part of the structure, is analysed under the effects of the loads it must sustain. Secondly, the critical sections are examined to check that they are capable of resisting the resulting actions. Using a simply supported beam as an example, the beam as a whole is first analysed to determine the distribution of bending moment under the applied loads and then the moment of resistance is calculated at the position of maximum bending.

Many national standards for the design of reinforced concrete structures now use a limit state approach. In simple terms, the values of the moments or shears at a particular section are calculated using factored characteristic loads; the resistance of the section is calculated using factored characteristic strengths. The situation is considered to be satisfactory if

$$\gamma_f \times \text{load effects} \leq \frac{structural\ resistance}{\gamma_m} \qquad [5.6]$$

where γ_f is the factor applied to characteristic loads and γ_m is the factor applied to characteristic strengths of material. The overall load factor γ_f is considered in many cases to be made up of three component factors such that:

$$\gamma_f = \gamma_{f1} \times \gamma_{f2} \times \gamma_{f3} \qquad [5.7]$$

γ_{f1} is the load variation factor and allows for uncertainty in assessing the characteristic loads;

γ_{f2} is the load combination and sensitivity factor which allows for the reduced likelihood of loads, acting in combination, all attaining their characteristic values simultaneously. It also takes account of the increased margin which is needed when some of the forces in a combination act in opposition; and

γ_{f3} is the structural performance factor and allows for inaccurate assessment of load effects, stress redistribution in the structure not taken account of in the calculations, the dimensional accuracy achieved in construction and the importance of the limit state under consideration.

In BS 8110,[5.6] the British code of practice for design of concrete structures, values of the individual components of γ_f are not given explicitly but the factor itself is quoted for various load combinations and limit states. For instance, for the ultimate limit state and a combination of dead and imposed load, the total load is calculated from:

$$1.4\ Gk + 1.6\ Qk \quad \text{(for adverse combinations of load)} \qquad [5.8]$$

or

$$1.0\ Gk + 0\ Qk \quad \text{(for beneficial combinations of load)} \qquad [5.9]$$

where Gk is the characteristic dead load; and
$\qquad Qk$ is the characteristic live load.

For this load combination and limit state γ_m is given as 1.15 for

Table 5.3 Value of the partial safety factors for load γ_f for various load combinations for the ultimate limit state given in Table 2.1 of BS 8110[5.6]

Load type	Load combination		
	1 Dead and imposed	2 Dead and wind	3 Dead, wind and imposed
Dead			
Adverse	1.4	1.4	1.2
Beneficial	1.0	1.0	1.2
Imposed			
Adverse	1.6		1.2
Beneficial	0		1.2
Earth and Water	1.4	1.4	1.2
Wind	—	1.4	1.2

Notes: 1. All load combinations may include earth and water pressure.
2. The 'adverse' partial factor is applied to loads that produce a more critical condition at the section considered. The 'beneficial' factor is applied to loads that produce a less critical condition.

Table 5.4 Components of γ_f for different loads and load combinations (after ISE[5.1])

	γ_{f1}		γ_{f2}	γ_{f3}
	Adverse	Beneficial		
1. Dead and imposed				
Dead	1.15	0.85	1.0	1.2
Imposed	1.35	0	2.0	1.2
Earth and water	1.15	1.15	1.0	1.2
2. Dead and wind				
Dead	1.3	0.85	0.9	1.2
Earth and water	1.3	0.85	0.9	1.2
Wind	1.15	0	1.0	1.2
3. Dead, wind and imposed				
Dead	1.15	1.15	0.9	1.2
Imposed	1.35	1.35	0.75	1.2
Earth and water	1.15	1.15	0.9	1.2
Wind	1.15	1.15	0.9	1.2

reinforcement and 1.50 for concrete in flexure or axial load. Values of γ_f for other load combinations given in BS 8110 are shown in Table 5.3.

The Institution of Structural Engineers Report[5.1] on the appraisal of existing structures indicates that a value of $\gamma_{f3} = 1.2$ was used in deriving the BS 8110 factors and this allows γ_{f1} and γ_{f2} to be calculated approximately for different components of load and load combinations as shown in Table 5.4.

Some of the uncertainties which are taken into account when deriving partial factors for the design of new structures are removed when carrying out assessments of existing structures. As an example the factors take account of tolerances which are inherent in any construction or manufacturing process. The load variation factor γ_{f1} on dead load takes into account the possibility that members could be constructed with slightly greater dimensions or density than the design values. Where dimensions and densities have been accurately established by survey, the partial factor can be reduced when carrying out structural assessment. The Institution of Structural Engineers report suggests that γ_{f1} could be reduced from 1.15 to 1.05 in these circumstances. In the case of thin slabs (100 mm or less) the report suggests that a figure of 1.10 may be appropriate. A similar reduction in γ_{f1} (to 1.05) is also suggested for loads from screens and partitions if their dimensions and densities have been measured.

The load combination and sensitivity factor γ_{f2} used in appraisals is the same as that used in design. The consideration that there is a reduced probability that various loads, acting together, will all attain their characteristic value at the same time is still valid in appraisals. A reduction in γ_{f3} from 1.2 to 1.15 or 1.05 is suggested in the Institution of Structural Engineers report depending on the structural importance of the member and the consequences of collapse. Inaccuracies of construction and analysis are two of the considerations taken into account when deriving values for γ_{f3}. If the assessment is based on measured dimensions, including eccentricities caused by constructional inaccuracy, and realistic or conservative assumptions are made about load transfer the report suggests:

γ_{f3} = 1.15 for primary members and for secondary members whose failure might cause loss of life and/or substantial damage
γ_{f3} = 1.05 for secondary elements whose failure will not lead to progressive collapse

It is also valid to consider reductions in the partial safety factors for strength of materials γ_m. As stated earlier, BS 8110 gives values of γ_m = 1.5 and 1.15 for concrete and steel, respectively. These values are said to take account of differences between actual and laboratory values, local weaknesses and inaccuracies in assessment of the resistance of sections. The Institution of Structural Engineers report suggests that consideration may be given to reducing γ_m for concrete to 1.25 in cases where:

1. Concrete strength has been ascertained by tests on cores taken from the structure.
2. The strength tests have been supplemented by ultrasonic pulse velocity or rebound hammer measurements to assess variability.
3. The failure mechanism is well understood and/or ductile.

A value of $\gamma_m = 1.35$ for concrete is suggested in the report for members where the failure mode is not clearly understood (e.g. shear) or where members may fail suddenly without warning (e.g. columns). There is not often a case for reducing the steel partial safety factor below 1.15, as it is unusual to be able to test samples of reinforcement from a number of representative members. However, if it is possible to obtain and test sufficient samples, the Institution of Structural Engineers report suggests a possible reduction to $\gamma_m = 1.05$, provided that:

1. The consistency of the mechanical properties of the reinforcement has been checked using non-destructive methods.
2. Measured effective depths are used in the calculations.
3. The full stress—strain curve for the reinforcement has been obtained from the tests and shows ductility and a reserve of strength beyond the yield point.

Currie[5.7] in a discussion of the Institution of Structural Engineers Report has shown how a logical approach to assessment of partial load factors applies in the particular example of a six storey *in situ* cast reinforced concrete framed structure infilled with brick panels and with composite beam and slab floors. Values of $\gamma_m = 1.25$ for concrete for members in flexure and $\gamma_m = 1.35$ for shear and column calculations are given in the particular example cited. However, it is suggested that the use of the 95 per cent confidence limit characteristic strength in assessment calculations may be unduly conservative in many circumstances. For instance, in order for a slab to fail under the action of a uniformly distributed load, all of the concrete along the yield lines must fail. In addition, there may be a degree of fixity at the edges and the slab cannot fail unless a considerable amount of concrete fails. Hence the behaviour of the slab is governed by the average strength of the concrete provided that the sample locations on which the strength is based were evenly distributed over the area.

A similar argument applies to T beams acting compositely, as adjacent beams share in supporting a load imposed on a single beam. Currie suggests that where significant concentrated loads are present or for shear calculations, a lower strength should be used in assessment calculations because the ability to distribute load would be critical and also to take account of the possibility of an area of weakness. It is suggested that the strength value corresponding to the 80 per cent confidence limit should be employed. A similar value is suggested for the case of columns where they tend to act in isolation or as part of the base frame, but the strength used in assessment could be further reduced if, for example, masonry infill panels provided an alternative load path. In the example quoted by Currie, the strength results were assessed separately for cores from different types of component, e.g. beams or columns.

High Alumina Cement Concrete Construction

A special subcommittee was set up by the Building Regulations Advisory Committee in the United Kingdom after the collapse of some roofs constructed from beams containing high alumina cement. In their report[5.2] they give recommendations on methods of assessment. The report suggests that some domestic buildings with HAC units should be exempt for appraisal. The exemption is for floors and roofs of standard factory-produced joists up to 250 mm depth in buildings consisting solely of houses, maisonettes and flats with the following provisos:

1. The building does not consist of more than four storeys.
2. In the case of roofs:
    (a) they are of joist and block or composite construction;
    (b) there is no access to the roof other than for maintenance;
    (c) the clear span does not exceed 6.5 m (8.5 m in the case of larger joists).
3. There is no persistent leakage or sustained heavy condensation.

For structures falling outside the above category, the report gives recommendations for concrete strength and partial safety factors for use in assessment. The recommendations are based on the use of BS CP110,[5.8] which was current at the time of the report. In the absence of actual data on concrete strength, a value of $21\ \mathrm{N\ mm^{-2}}$ is suggested along with the partial safety factors shown in Table 5.5.

During the Committee's investigation, it had been found that in some cases the tops of joists had been trowelled smooth by hand using a sand/cement mortar. This process could have led to concrete near the top surface having a high water/cement ratio and which was deficient in coarse aggregate. For this reason, the committee recommended that γ_m for concrete should be taken as 1.5 for isolated beams. A figure of

Table 5.5 Criteria for appraisal of high alumina cement concrete beams in non-composite construction (after the Building Regulations Advisory Committee[5.2])

| | Concrete strength Fcu (N mm^{-2}) | Beam performance factor | Partial safety factors | | | |
| | | | Material (γ_m) | | Loading | |
Use			Concrete	Steel	Dead	Live
Joist and hollow block construction	21	1.3	1.0	1.0	1.4†	1.6
Isolated beams or purlins	21	1.3	1.5*	1.0	1.4†	1.6

Notes: Criteria are for standard factory-produced cross-joists up to 250 mm deep manufactured on a long term production line with consistent dimensional and quality control.
*Where it has been verified that the top surface had not been trowelled smooth a value of $\gamma = 1.0$ may be used.
†Where actual dead load is assessed the factor may be reduced to 1.2 for floors and 1.3 for roofs.

$\gamma_m = 1.0$ can be used if it has been verified that the top surfaces of the joists had not been trowelled smooth.

The beam performance factors given in Table 5.5 are based on results of tests carried out on joists. The joists were analysed using the partial factors shown and the results compared with the actual moments from the tests. By this means, it was possible to derive acceptable lower bound values for performance factors for use in appraisals.

Load Tests

Load test results, whether they be in the form of deflection measurements or strain measurements, first need to be corrected for effects such as bearing movements and changes in environmental conditions during the test before they are assessed for compliance.

BS 8110[5.6] gives guidelines on the assessment of load tests in which it states that the main objective is to compare measured performance with that predicted by calculation. For the test, the loading is given as the greater of

(a) the characteristic dead load plus 1.25 times the characteristic imposed load or
(b) 1.125 times the sum of the characteristic dead and imposed loads.

Two cycles of loading and unloading are applied.

Three specific criteria for acceptable performance are given:

1. Initial deflections in accordance with design requirements ($L/500$).
2. Where significant deflections have occurred, the percentage recovery of deflection after the second cycle of loading should be at least equal to the percentage recovery after the first cycle and also be at least 75 per cent.

The BS 8110 requirements are similar to those given in the previous code CP110[5.8] but the latter document gives a little more detail. Under CP110 the loads are similar to those given in BS 8110 but for the ultimate limit state the load is maintained for a period of 24 hours. Deflection is measured immediately after application of load and the requirement is for a deflection of less than 1/500 of the effective span. If the maximum deflection in millimetres during the 24 hours under load is less than $40 \times L^2/h$ (L = effective span in metres and h = overall depth of construction in millimetres), it is not necessary to measure recovery. Where recovery is measured, it is required to be 75 per cent within 24 hours of removal of the test load. If this value is not achieved, the load test is repeated and the structure is considered to have failed the test if the recovery after the second loading is not at least 75 per cent of the deflection under the second loading.

Chloride Contents

As discussed in Chapter 2, chloride may have been present in the concrete at the time of mixing or may have penetrated from some external source during service. To determine which is the case investigations usually involve determination of chloride concentrations at various depths from the surface of a structure. The results are plotted to show the variation in chloride with depth. Two cases are shown in Fig. 5.7. In Fig. 5.7(a)

Fig. 5.7 Graphs of chloride concentration against distance from surface. (a) Where chloride was included in the mix; (b) where chloride has entered the concrete from an external source

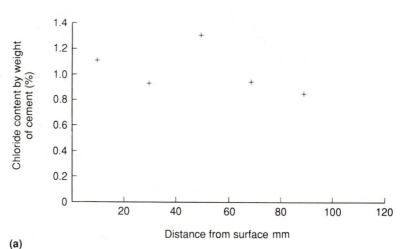

(a)

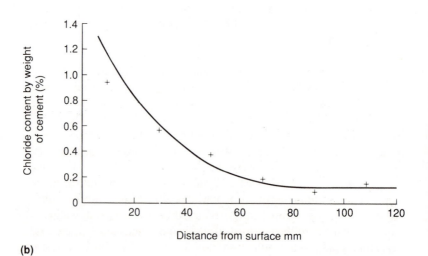

(b)

the chloride concentration is high and reasonably uniform illustrating the situation where chlorides were added to the mix or were present in the ingredients. Figure 5.7(b) illustrates the case where chlorides have penetrated the hardened concrete. There are high concentrations near the surface and the values decrease with increasing distance from the surface until they reach the background levels which were present in the concrete at the time of construction.

If the chloride has entered by a diffusion mechanism obeying Fick's law, it is possible to calculate a diffusion coefficient from results which indicate a profile. This can be used to predict the profile at some time in the future and may assist in making assessment of the time period before corrosion is likely to occur if critical chloride concentrations can be established. A simpler approach, also based on the application of Fick's law, is to use the fact that the depth at which a particular concentration occurs is proportional to the square root of the time of exposure. As an example, if a critical concentration occurs at depth 15 mm after an exposure period of 5 years, the critical concentration will not reach a depth of 30 mm until a total exposure period of 20 years has elapsed. This relationship can be used to assess the time when the critical chloride concentration is likely to occur at the position of the reinforcement. Caution needs to be exercised in using the relationship as it applies in theory only when a solution of constant concentration is permanently in contact with the outside surface of saturated concrete.

Funahashi[5.9] has suggested a finite difference calculation method for predicting chloride profiles after varying times of exposure. The method is based on Fick's law of diffusion and a method is given for checking whether a chloride profile has been produced by diffusion. The results are plotted on a probability scale as shown in Fig. 5.8. If the points lie on a straight line, chloride has entered the concrete by a diffusion mechanism obeying Fick's law.

Cracking and spalling due to corrosion do not take place immediately a critical value is reached at the reinforcement. It is sometimes useful to think of the time to significant corrosion damage $t(t)$ as the sum of two separate intervals as shown in Fig. 5.9 and expressed as follows:

$$t(t) = t(i) + t(p) \qquad [5.10]$$

where $t(i)$ is time to initiation of corrosion; and

$t(p)$ is the time for corrosion to progress to such an extent that significant damage occurs.

The critical chloride concentration which triggers corrosion is the subject of some debate. In the case where chloride had its origin at the time of construction, it is usual to compare the test results with the chloride limits in national standards. British Standard BS 8110[5.6] specifies a limit of 0.4 weight per cent of cement for ordinary Portland

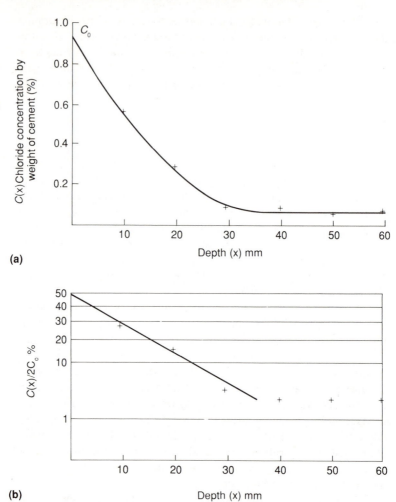

Fig. 5.8 Chloride profiles plotted. (a) with natural chloride concentration axis; (b) with chloride concentration axis to probability scale. Note. $C(x)$ is the chloride concentration at depth x, C_o is the chloride concentration at the surface

cement concrete and 0.2 weight per cent of cement for sulphate resisting cement concrete. The Building Research Establishment in the United Kingdom have suggested[5.10] that the risk of corrosion may be only moderate for chloride concentrations up to 1.0 weight per cent of cement in dry conditions if carbonation has not penetrated to the position of the reinforcement. This again is specifically related to the situation where chloride was introduced at the time of concrete production.

It can be argued that where chloride has penetrated the structure during service the construction limits mentioned above do not apply. This is because chloride present in the mix takes part in the hydration reactions and becomes chemically bound into the cement matrix. It is probable the depassivation of the reinforcement and consequent corrosion is more

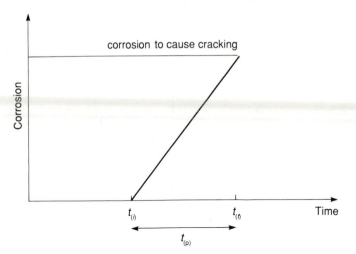

Fig. 5.9
Conceptual mode
to show
development of
corrosion with time

directly affected by chloride in the pore solution. Chloride bound up in the hydration products may be released only slowly into the pore solution until there is a balance between the dissolved and the combined chloride. Hence the critical chloride concentration is lower in the case where it enters from an external source and is less likely to be chemically bound to the products of hydration.

In the United States two methods of analysis for chloride content are in common use. In some instances water extraction of the sample is used. In others acid extraction is employed to determine the 'total' chloride content. A report on highway bridges[5.11] quoting earlier work by Lewis[5.12] suggests that the corrosion threshold value is 0.15 per cent of acid-soluble chloride by weight of cement. Work mentioned in the report had shown that water-soluble chloride is on average 75−80 per cent of acid soluble chloride in hardened concrete subject to de-icing salts. This leads to a threshold value of 0.12 per cent water-soluble chloride by weight of cement. As the test for soluble chloride in concrete is difficult to control and takes longer than the test for total chloride, the report suggests that the latter method is used as routine. Only if the reported values exceed 0.15 weight per cent of cement should a check be carried out using the water-soluble method.

Carbonation

Carbonation depths in themselves are not very useful except that they may give a general indication of the quality of the concrete. The depths need to be considered in relation to the cover to the reinforcement. One approach is to plot the carbonation depth results on a histogram with the covermeter survey results, as shown in Fig. 5.10(a). There is clearly

Fig. 5.10
Histograms
showing
(a) distribution of
cover to
reinforcement and
carbonation depth
measurement
results; (b) results
from (a) plotted as
a ratio

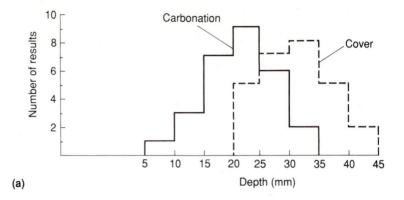

(a)

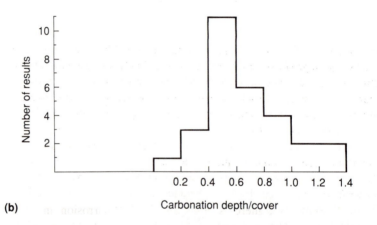

(b)

an overlap in the two ranges shown and the potential for corrosion may exist if some reinforcement lies within the carbonated zone. A more useful procedure is to plot the ratio of carbonation depth to cover at each position. This has been done in Fig. 5.10(b). It can be seen that at 4 out of 28 locations, carbonation has reached the reinforcement. If the results are representative this means that the reinforcement may be at risk of corrosion at over 15 per cent of the surface area of the structure. Actual occurrence of corrosion will depend on other factors such as availability of moisture, and the situation needs to be checked by breaking out selectively at locations that include a range of carbonation depth to cover ratios.

Diagrams of the type shown in Fig. 5.10(b) can also be used to assess the percentage of reinforcement likely to be at risk of corrosion at some time in the future. If the data in Fig. 5.10(b) represents the situation after 15 years, the relationship between the square root of time and carbonation depth means that carbonation depths will have increased by a factor of $\sqrt{2} = 1.41$ when the exposure time is 30 years. Therefore the carbonation depth will exceed the cover to the reinforcement after 30 years at any location which has a carbonation depth to cover ratio in excess of $1/\sqrt{2} = 0.71$ at present. In the case shown in Fig, 5.8, at least 30 per cent of the reinforcement will be in this condition after 30 years and will be at risk of corrosion.

Carbonation and Chlorides

The situation where medium to high chloride concentration coexist with carbonation depths close to the reinforcement is even more difficult to assess. Some guidance has been given by the Building Research Establishment in the United Kingdom[5.10] for the case when the chloride was included in the original mix. Three ranges of chloride concentration are considered in association with carbonation depths less than or greater than the cover to reinforcement, and two broadly defined exposure conditions. Stated briefly, the guidance given is that:

1. If the chloride content is less than 0.4 per cent by weight of cement and the carbonation depth is less than the cover to reinforcement, there is a low risk of corrosion in all environmental conditions.
2. If the chloride content is less than 0.4 per cent by weight of cement and the carbonation depth is greater than the cover to reinforcement, there is a moderate risk of corrosion in damp conditions.
3. If the chloride content is in the medium range 0.4 to 1.0 per cent by weight of cement and the carbonation depth is less than the cover to reinforcement, there is moderate risk of corrosion in damp conditions.
4. If the chloride content is in the medium range 0.4 to 1.0 per cent by weight of cement and the carbonation depth is greater than the cover to reinforcement, there is a high risk of corrosion which is exacerbated by damp conditions.
5. If the chloride content is in excess of 1.0 per cent by weight of cement there is a high risk of corrosion irrespective of the carbonation depth. The risk is increased in damp conditions.

These conditions are very useful in assessing the risk of corrosion from the results of tests. One way of using them is to produce a scatter plot, as shown in Fig. 5.11, with a risk of corrosion assigned to six regions dependent upon chloride content and carbonation depth according to the categories given above. This allows a quick appreciation to be gained of the proportion of samples in the various categories and a rapid assessment of the overall risk. Separate identification of those locations likely to encounter damp conditions is also useful. It is worth breaking

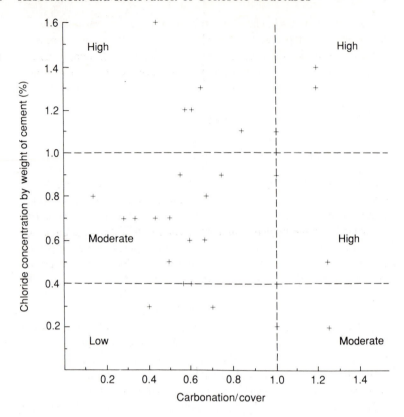

out some areas selectively to check the conditions of the reinforcement so that the categories can be related to the actual condition of the structure rather than the risk.

The position of the vertical line which divides the categories according to carbonation depth could be adjusted as discussed previously for Fig. 5.10, to allow an approximate prediction of the risk at some future time to be made if conditions remain constant.

It should be noted that the above categories are those associated with chloride incorporated in the concrete at the time of construction. As chloride ingress mechanisms and the resulting corrosion threshold values become better understood, it may be possible to utilize diagrams like Fig. 5.11 but with the vertical axis also expressed as a ratio. In this case, it would be the ratio of the depth at which the chloride threshold value occurs and the cover to reinforcement.

Morinaga,[5.13] reported by Choong Kog,[5.14] has given a conceptual degradation model for the performance of concrete affected by chlorides and carbonation (Fig. 5.12). He suggests that cracking of concrete will

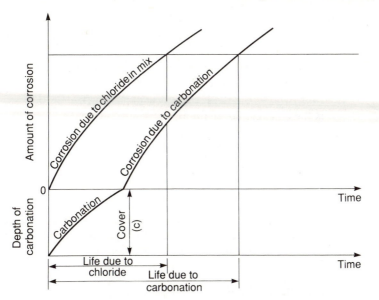

Fig. 5.12
Conceptual model
to show
development of
corrosion with time
(after Choong
Kog[5.14])

first occur when there is a certain quantity of corrosion on the reinforcement. The amount is given by:

$$Q_{cr} = 0.602d\,(1 + 2c/d)^{0.85} \qquad\qquad [5.11]$$

where Q_{cr} is the critical mass of corrosion product ($\times 10^{-4}\,\mathrm{g\,cm^{-2}}$);
    c is the cover to reinforcement (mm); and
    d is the diameter of reinforcing bars (mm).
The time for cracking to take place is given by:

$$t_2 = Q_{cr}/q \qquad\qquad [5.12]$$

where q is the corrosion rate due to either carbonation or chlorides in gram per day.

The case shown by Choong Kog relates to the situation where chlorides were present in the concrete at the time of mixing. In the case of carbonation (or the case of chlorides entering from an external source), the total time between construction and cracking is given by:

$$t_r = t_1 + t_2 \qquad\qquad [5.13]$$

where t_1 is the time taken for the carbonation front to reach the reinforcement when corrosion starts.

Morinaga suggests that t_1 can be calculated for different environmental conditions and concretes with different water/cement ratios from the equation:

$$t_1\,(\text{days}) = \{c/(A \times K_e \times K_c)\}^2 \qquad\qquad [5.14]$$

where A is a coefficient of carbonation velocity ($= 1.09\sqrt{C}$);

K_e is a factor depending on relative humidity and temperature ($=$ $1.391 - 0.0174rh + 0.0217T$);

K_c is a factor depending on the water/cement ratio ($= 4.6 \, (w/c) - 1.76$ when $w/c < 0.6$ $= 4.9 \, (w/c - 0.25)/\{1.15 + 3(w/c)\}^{0.5}$ for $w/c > 0.6$);

c is the concrete cover (mm);

rh is the ambient relative humidity (per cent)

T is the ambient temperature (°C);

w/c is the water/cement ratio; and

C is the concentration of carbon dioxide in the atmosphere (per cent).

Similar equations are given for calculating the rate of corrosion due to chlorides. The life of the structure (until first cracking) is the lesser of the values calculated for carbonation or chloride.

Sulphate Contents

Sulphate is added to cement during the manufacturing process to control the rate of set. National standards for cement usually include a limitation on the total sulphate content. British Standard 12[5.15] specifies a limit of 3.5 per cent expressed as SO_3 when the tricalcium aluminate content is greater than 3.5 per cent and 2.5 per cent if the tricalcium aluminate content is less than 3.5 per cent. ASTM C 150[5.16] specifies a limit of 3.0 per cent if the tricalcium aluminate content is 8 per cent or less and 3.5 per cent if the tricalcium aluminate content is greater than 8 per cent for Type I Cement. For Type V (sulphate-resisting cement) the limit is 2.3 per cent.

As a consequence of the above, concrete mixes usually have a sulphate content in the range 2.0–3.5 per cent by weight of cement arising from the cement. British Standard 8110[5.6] sets a limit of 4.0 per cent SO_3 by weight of cement for concrete mixes, and this value is also sometimes used when assessing the results from concrete investigations. However, if results are marginally above this figure and there are no visual external signs of deterioration, it might be appropriate to undertake petrographic analysis of a sample from the concrete. This should enable the form in which the sulphate is occurring within the concrete to be identified and particularly, whether there has been any formation of ettringite (calcium sulphoaluminate) likely to cause deleterious expansive reactions.

Half-cell Potential Results

Half-cell potential results are often assessed by producing contour maps of the structure surface, as shown in Fig. 5.13 The significance of half-

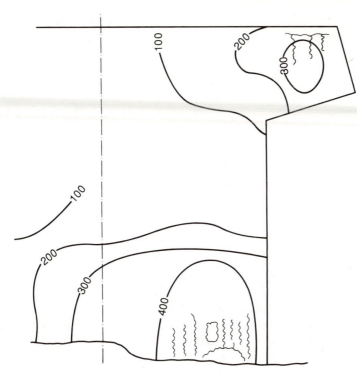

Table 5.6 Guidance on interpretation of results from half-cell surveys given in ASTM C 876[5.17]

Half-cell potential more positive than −0.2 V
There is a greater than 90 per cent probability that no reinforcing steel corrosion is occurring in that area at the time of measurement.

Half-cell potential in the range −0.2 V to −0.35 V
Corrosion activity in the area is uncertain.

Half-cell potentials more negative than −0.35 V
There is a greater than 90 per cent probability that reinforcing steel corrosion is occurring in that area at that time.

---

cell readings and their relationship to potential for corrosion is well documented and a statement on interpretation of results is included in ASTM C 876.[5.17] The advice on interpretation from ASTM C 876 is summarized in Table 5.6, which relates to values obtained using a copper/copper sulphate half-cell.

The advice on interpretation is based on laboratory testing and also experience from testing of highway bridges in the United States that were contaminated with road salts.[5.18] The validity of these interpretations in other situations has been questioned,[5.19] particularly in the case of saturated concrete where there may be negative potentials of high magnitude but insufficient oxygen to sustain corrosion. It is suggested that steep gradients of potential may be good indications of corrosion activity. It has also been noted that plots of potential contours allow the sites of anodic areas to be identified.[5.20]. These sites are possible areas of localized corrosion.

Alkali–Silica Reactivity

One of the results of alkali–silica reactivity is an extensive network of interconnected cracks throughout the mass of the concrete, as described in Chapter 2. The cracks represent a pathway by which oxygen and moisture can gain access to the reinforcement and this aspect needs to be taken into account in any assessment of future performance. The cracks also result in changes in the physical properties of specimens retrieved from affected members.

The Institution of Structural Engineers[5.21] has suggested the use of an expansion index when assessing structures suffering from alkali–silica reaction. The index takes into account the current expansion and potential additional expansion of the structure for which the test methods are described in Chapter 4. It is suggested that the laboratory expansion test is carried out at either 5, 13 or 20 °C to model the conditions in the structure as closely as possible.

In order to obtain results within a reasonable time period four cores are tested and the greatest expansion on any core after six months is used. Results are plotted on a time-scale and extrapolated to two years. The predicted value after two years is referred to as the potential additional expansion. The total expansion is calculated by adding the current free expansion (measured by summing the widths of cracks crossed by lines drawn on the surface of the structure) and the potential additional expansion. The expansion index is determined by plotting values for estimated current expansive strain and estimated total expansive strain on a graph, as shown in Fig. 5.14. The plotting area is divided into five regions which are given the roman numerals I to V and each of which corresponds to a different expansion index.

The Institution of Structural Engineers report suggests that the expansion index can be used to provide a basis for the future management of affected structures. However, the report emphasizes that it is not possible at the present time to give firm recommendations which cover the full range of possible expansions due to alkali reactivity, and also that some of the current recommendations may need to be revised in the light of further experience. It is understood that a revised procedure is in the course of preparation. Performance is related to the degree of

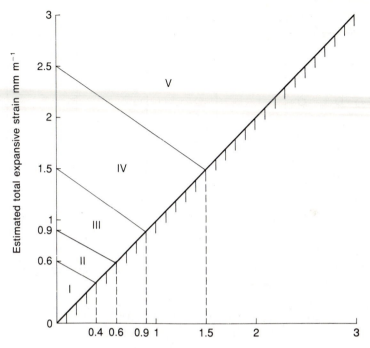

Fig. 5.14 Chart to determine expansion index as suggested by the Institution of Structural Engineers[5.21]

containment that the reinforcement is likely to provide to the expansive forces generated during the reaction. For the purposes of appraisal, members are divided into three classes related to the way in which the reinforcement is detailed:

Class 1   Members which contain a three-dimensional cage of very well anchored reinforcement.

Class 2   Members which contain a three-dimensional cage of conventionally anchored reinforcement.

Class 3   Members which contain a one- or two-dimensional cage of reinforcement, or a three-dimensional cage that is inadequately anchored.

The classes are illustrated, using the example of a wall, in Fig. 5.15. It is suggested in the report that two structural appraisals should be undertaken — one using the expansion at the time of the investigation and one using the predicted expansion at some time in the future. A summary of the suggested appraisal procedures is given in Table 5.7.

A maintenance strategy for structures affected by alkali—silica reactivity is also suggested in the Institution of Structural Engineers

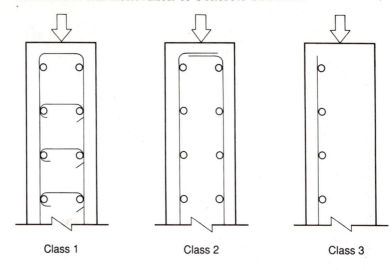

Fig. 5.15 Detailing classes suggested by the Institution of Structural Engineers for the management of structures affected by alkali reactivity[5.21]

Class 1   Class 2   Class 3

report. Each member of the structure is assigned a structural severity rating depending on:

(a) site environment (dry, intermediate, wet);
(b) reinforcement detailing class (as described above);

Table 5.7 Guidance on structural appraisal requirements related to class of reinforcement detailing and total expansion (after Doran[5.21])

Total expansion	Class 1	Class 2	Class 3	Comments
Less than 0.6 mm m⁻¹	OK	OK	Assess for adequacy	Any shortfall in structural strength in Class 1 or 2 members may be assumed to be covered adequately by the normal safety factors provided that the structure has been properly designed.
0.6–0.9 mm m⁻¹		OK	Assess for adequacy	Comment above applies to Class 1 members within this range of total expansion. Class 2 and 3 members should be assessed with care, paying particular attention to shear and load which may be adversely affected by the reduction in tensile strength. There may be severe local problems or corrosion or frost damage if cracks exceed 1 mm in width.
0.9–1.5 mm m⁻¹		Detailed appraisal for all classes		Mild steel reinforcement may yield. Full consideration needs to be given to possible reductions in bending and compression capacities as well as shear and bond.
1.5–2.5 mm m⁻¹		Detailed appraisal for all classes		High yield steel may yield. Comments for 0.9–1.5 mm m⁻¹ total expansion range apply.
Greater than 2.5 mm m⁻¹		Special study, testing, appraisal and monitoring for all classes		Load tests may be appropriate.

(c) expansion index;
(d) consequence of further deterioration (slight or significant).

The 'slight' category under consequence of further deterioration is appropriate to these members where the consequences of structural failure are either not serious or are localized to the extent that a serious situation is not anticipated. The 'significant' category under consequence of further deterioration is appropriate to those members where there is a risk to life and limb, or a significant risk of serious damage to property.

The five categories of structural severity range from very mild to very severe through mild, serious and severe. All members with an expansion index of I lie in the very mild category, except that members in a wet environment detailing Class 3 and a 'significant' consequence of failure rating are classified as mild. At the other end of the scale, all members with an expansion index of V are placed in the very severe category except that members in a dry environment with Class 1 detailing and 'slight' consequence of failure rating are classified as severe.

An appropriate course of action for each of the structural severity ratings is given in the report. These range from organization of routine maintenance inspections to report on the development of cracking for the 'very mild' rating to detailed specialist study, and the possibility of load testing and measures to reduce the rate of deterioration for the 'very severe' rating. The full range of management procedures is summarized in Table 5.8.

As would be expected from a mechanism which causes cracking, there is a much greater effect on tensile strength than compressive strength. The modulus of elasticity and ultrasonic pulse velocity are also reduced. It would be anticipated that a reduction in concrete strength would result in a commensurate reduction in the load-carrying capacity of affected members. In particular, the lower tensile strength could be expected to reduce the bond between reinforcement and concrete and also the shear strength of a section.

In order to quantify the effects of alkali—silica reaction on the integrity of structures, the Building Research Establishment has carried out tests[5.22] on a number of full-scale beam specimens. The types of members chosen were pretensioned I-beams of a kind used in bridge construction and some buildings. Members of this type were chosen firstly because, as stated above, loss of bond was a principal concern and bond performance could be assessed in terms of loss of prestress. A second reason was that high strength concrete with high cement contents is normally used in prestressed concrete. This means that they are more likely to have increased alkali contents and are therefore potentially susceptible to alkali—silica reaction. Two sets of beams were investigated, one with shear reinforcement and the other without.

Table 5.8 Management of structures affected by alkali−silica reaction related to structural severity rating (after Doran[5.21])

Structural severity rating	Management procedure
Very mild	Alkali−silica reactivity of no current significance. Carry out routine maintenance inspections. Report systematically on the development of cracking or deformation.
Mild	Install Demec studs across selected cracks and also in triangular formation where shear movement is to be checked — read every 6 months. Take cores and test for expansion over a two-year period. Compare laboratory and field data annually. Carry out three-yearly engineering inspection. Consider reducing access of water to the structure.
Serious	Monitor cracks. Consider other forms of instrumentation. Take cores and test for expansion. Carry out annual engineering inspections. Reduce access of water to the structure where possible.
Severe	Extensive and carefully planned instrumentation of the structure. Extensive laboratory testing. Full investigation of structural detailing. Remedial or strengthening work may be needed. Load restriction may be necessary. Carry out engineering inspections every 3 months. Access of water to the structure to be reduced.
Very severe	Immediate action required. Detailed specialist study required which may include load testing. Measures may be required to reduce the rate of deterioration.

The beams were constructed from a concrete mix containing reactive Thames Valley sand and were cured in water tanks at 38 °C to speed up the reaction. After 5 months expansion was complete and sections of the beams were tested for flexural and shear capacity, and loss of prestress. Flexural capacity was determined under four-point loading with a simply supported span of 5 m and symmetrically placed point loads 0.5 m apart. Shear capacity was determined under a single point load applied at a distance of twice the section depth from the centre of one support on a 2.5 m span. The remaining level of prestress was determined by measuring tendon length, debonding, and remeasuring and making allowances for transmission length.

It was found from the tests that there was no significant reduction in prestress or flexural capacity. A reduction in shear capacity of about 20 per cent was found at the time of first cracking in beams both with and without shear reinforcement. However, as expansion increased and the resistance from the links created a vertical prestress, the beams with shear reinforcement regained their capacity. This effect was absent in the beams without shear reinforcement.

Hobbs[5.23] has reported tests in Japan, South Africa and Denmark which included both work on laboratory specimens and loading of actual structures affected by alkali—silica reactivity. In the case of testing on actual structures, the capacity was found to be adequate even in the deteriorated condition. In the case of laboratory studies, the beams affected by alkali—silica reaction were found to have capacities which were similar to control specimens.

References

5.1   Institution of Structural Engineers 1980 *Appraisal of Existing Structures* The Institution of Structural Engineers, London

5.2   Building Regulations Advisory Committee 1975 *Report by Sub-committee P (High Alumina Cement Concrete)* Department of the Environment, London

5.3   Hind O K R, Bergstrom W R 1985 Statistical evaluation of the in-place compressive strength of concrete. *Concrete International* **Feb:** 44—8

5.4   Watkins R A M, McNicholl D P 1990 Statistics applied to the analysis of test data from low-strength concrete cores. *The Structural Engineer* **68 (16):** 327—32

5.5   Bartlett F M, Sexsmith R G 1991 Bayesian technique for evaluation of material strengths in existing bridges. *ACI Materials Journal* **Mar—Apr:** 164—9

5.6   British Standards Institution 1985 *The Structural Use of Concrete* BS 8110, The British Standards Institution, London

5.7   Currie R J 1990 Towards more realistic structural evaluation. *The Structural Engineer* **68 (12):** 223—8

5.8   British Standards Institution 1972 *The Structural Use of Concrete* CP110, The British Standards Institution, London

5.9   Funahashi M 1990 Predicting corrosion-free service life of a concrete structure in a chloride environment. *ACI Materials Journal* **Nov—Dec:** 581—7

5.10   Building Research Establishment 1982 *The Durability of Steel in Concrete: Part 2. Diagnosis and Assessment of Corrosion-cracked Concrete* Digest 264, The Building Research Establishment, Garston, United Kingdom

5.11   National Co-operative Highway Research Program 1979 *Durability of Concrete Bridge Decks* Synthesis of highway practice 57, Transportation Research Board, National Research Council, Washington

5.12   Lewis D A 1962 Some aspects of the corrosion of steel in concrete. *Proceedings of the First International Congress on Metallic Corrosion*, London

5.13   Morinaga S 1986 *Prediction of Service Lives of Reinforced Concrete Buildings Based on Rate of Corrosion of Reinforcing Steel* Special Report of the Institute of Technology, Skimizu Corporation, Japan

5.14 Choong Kog Y 1989 Appraisal of an existing three-storey concrete building. *The Structural Engineer* **65 (20):** 360–3

5.15 British Standards Institution 1989 *British Standard Specification for Portland Cement* BS 12, The British Standards Institution, London

5.16 American Society for Testing and Materials *Standard Specification for Portland Cement* ASTM C 150, The American Society for Testing and Materials, Philadelphia

5.17 American Society for Testing and Materials *Standard Test Method of Half-cell Potentials of Uncoated Reinforcing Steel in Concrete* ASTM C 876, The American Society for Testing and Materials, Philadelphia

5.18 Van Daveer J R 1975 Techniques for evaluating reinforced concrete bridge decks. *Journal of the American Concrete Institute* **72:** 697–703

5.19 Concrete Society 1989 *Cathodic Protection for Reinforced Concrete* Technical Report No 36, The Concrete Society, Slough

5.20 Vassie P R 1984 Reinforcement corrosion and the durability of reinforced concrete bridge *Proceedings of the Institution of Civil Engineers, Part 1* **76 Aug:** 713–23

5.21 Doran D K (Chairman) 1988 *Structural Effects of Alkali–Silica Reaction — Interim Technical Guidance on Appraisal of Existing Structures* Report of an *ad hoc* committee of the Institution of Structural Engineers, London

5.22 Clayton N, Currie R J, Moss R M 1990 The effects of alkali–silica reaction on the strength of prestressed concrete beams. *The Structural Engineer* **68 (15):** 287–92

5.23 Hobbs D W 1988 *Alkali–Silica Reaction in Concrete* Thomas Telford Ltd, London

6   Repair and Renovation Techniques

Preliminary Considerations

The choice of method of repair of a concrete structure is not often straightforward. There will usually be at least two possible approaches which are technically acceptable and the ultimate choice depends on other considerations. Other factors which need to be taken into account are:

1. Cause of damage and the results of the assessments described in Chapter 5.
2. Future life requirement of the structure.
3. The overall quantity of repairs and the size of individual repairs.
4. Access.
5. Requirement for continued use of the structure during repair and the time available for repair.
6. Relative costs.
7. Client requirements including future maintenance and economic considerations.

Some examples of how these factors influence choice of repair method may help to illustrate the decision-making process. Fire usually causes damage to large areas of concrete on beams, columns and soffits of slabs. The reinforcement may be affected. Sprayed concrete is often chosen to repair fire-damaged structures because it can be applied quickly and economically to the large areas involved and new reinforcement can be incorporated relatively easily. Carbonation and the resulting corrosion damage on the façade of a building usually results in a requirement for relatively small isolated repairs. Patch repair using cement-based or polymer mortars is often used in these circumstances. In the case of chloride-contaminated reinforced concrete, depending on the degree of contamination, there may be an option between cutting out and replacing whole members and undertaking patch repairs and monitoring performance if only a short period of further life is required for the

:ture. Repairs on structures associated with industrial processes may
: to be carried out during brief maintenance shut-down periods.
ayed concrete may be the only viable method of placing the volumes
concrete required in the available time.

Common techniques used in repair and renovation include:

(a) sprayed concrete,
(b) breaking out and recasting of individual members or regions of
     structures,
(c) patch repair,
(d) crack injection.

Access for Repair

Good access for operatives, supervisors and inspectors is an essential
feature of a successful repair contract. In many cases full scaffolding
is desirable as it gives the greatest flexibility in operations and the best
working platform. Scaffolding also affords the opportunity to provide
full shelter against the elements at the workplace and to control the
environment in which repair materials are being used. However, it can
be an expensive option in direct cost terms, and owners of buildings may
not wish to have the full façade of their structure covered with scaffolding.

Cradles and self-elevating platforms are also widely used. Many tall
buildings are provided with permanently fixed cradles used for window
cleaning and maintenance. They may only be suitable for light loading
and may not have the capacity to carry two or three men and their
equipment. The possibility of lost time due to cradle breakdown also
needs to be carefully considered where a contractor takes over the existing
cradles on a structure.

Self-elevating platforms consist of one or two towers, which may be
free-standing or bolted to the structure, and a working staging with guard
rails. The platform is powered by electrical motors which permit the
working staging to climb up the towers. Self-elevating platforms provide
a more rigid working platform than cradles and may also have a greater
load-carrying capacity. However, both cradles and platforms may be
a little inconvenient if the repair procedures require a variety of operations
using different equipment and multiple visits to the repair site are
necessary. Their use also means that work has to be suspended for a
time if the contract supervisor wishes to inspect the area.

Cleaning

Cleaning of the concrete surfaces is an essential element of many
rehabilitation contracts. It may be necessary to clean before applying

a decorative or protective coating, but it is also advisable before surveying to locate defects as dirt may obscure important features such as cracks. Many cleaning procedures are available and the choice will depend on the condition of the substrate and the degree and nature of contamination.

In the majority of cases, high pressure water jetting will produce the desired standard, but in cases of heavy pollution or ingrained dirt it may be found necessary to use wet sand blasting or manual scrubbing with detergents.

Survey

The survey carried out before the rehabilitation contract to determine the underlying causes of the problems is unlikely to have covered the whole of the structure and located all of the defects. Once access has been provided and cleaning has been completed, a comprehensive survey can be put in hand so that all of the areas requiring repair can be identified and the appropriate repair procedures can be implemented.

Sprayed Concrete

Sprayed concrete is produced by a process in which a stream of material is projected at high velocity into the required position. The process produces a dense concrete and no additional compaction is required. The use of sprayed concrete in construction was first developed in the United States from a technique used for producing large models of animals from plaster sprayed onto a wire framework. The equipment for sprayed concrete was patented in the United States under the name of the Cement Gun and the material was called Gunite. The word 'gunite' is not now restricted by patents and is used commonly by engineers in the United Kingdom.

The Concrete Society has produced a code of practice,[6.1] specification[6.2] and guidance notes on a method of measurement[6.3] for sprayed concrete. In these documents, gunite is defined as a sprayed concrete with a maximum aggregate size of less than 10 mm, whereas shotcrete is defined as a sprayed concrete where the maximum aggregate size is 10 mm or greater. In the United States, the terminology is different and shotcrete is defined by the Americann Concrete Institute as mortar or concrete conveyed through a hose and pneumatically projected at high velocity onto a surface.

Dry Mix Process

There are two separate basic methods of producing sprayed concrete: these are the dry mix and wet mix processes. In the dry mix process,

Fig. 6.1 Sprayed
concrete — dry
mix system

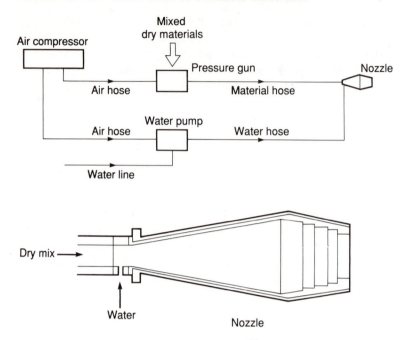

illustrated in Fig. 6.1, the cement, sand and aggregate are mixed together
in the dry state. The dry ingredients are transferred to the gun where
they are fed into a chamber and blasted by high pressure air down a
hose to the nozzle. Water is injected at the nozzle to produce a stream
of concrete of the desired consistency. The amount of water added can
be altered by adjusting a valve under the control of the operator or
nozzleman.

Wet Mix Process

In the wet mix system, illustrated in Fig. 6.2, all of the ingredients
including water are mixed together in the conventional manner. The mix
is placed in a concrete pump which causes it to travel at relatively low
speed along the delivery hose. At the nozzle, compressed air is introduced
into the stream and projects the mix into position at high velocity.

   In the United Kingdom, the original dry mix process has been used
more commonly than the wet mix process. Each of the processes has
its own advantages and disadvantages, and the choice of process may
well depend on the particular circumstances of the application. In the
wet mix process the mixed material is supplied from normal batching
operations and the proportioning can be carefully controlled. The amount
of water added in the dry mix process is controlled by the nozzleman
and may be variable. However, the range of water/cement ratios which

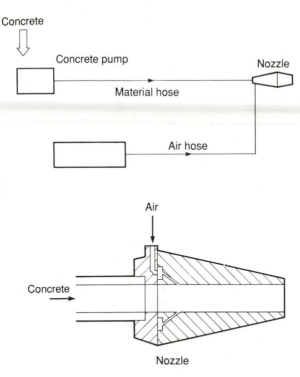

Fig. 6.2 Sprayed concrete — wet mix system

will result in good compaction and little or no slump is limited, and it should soon become evident if too much or too little water is being added. The dry process produces more dust and rebound than the wet process. The wet process, therefore, tends to be favoured for work in confined spaces such as tunnels and in environmentally sensitive areas.

The dry process is appropriate where sprayed concrete is to be used intermittently because no water is added to the mix until just before placing. Spraying can be stopped and started at will. The wet mix process is more suited for continuous spraying and has higher outputs. It is often used where large volumes of material have to be placed.

Materials and Application

It is usual to specify sprayed concrete by strength grade and to leave detailed mix design to the contractor. Cement, sand and aggregate should comply with relevant standards for concrete. Coarse sand with a 5 mm-down grading is usually employed. It has been noted[6.4] that a moisture content of 5–8 per cent in the sand is desirable as this level guards against segregation of the dry materials, but is not sufficient to cause blocking of the equipment. Most commonly-used mixes for the dry mix process have an aggregate cement ratio in the region of $3\frac{1}{2}:1$, but mixes in the range $2:1$ to $5:1$ have been used successfully. When the mix is shot into position some of the mixture rebounds. The rebound material does not contain as much cement as the shot material and the sprayed concrete in-place will usually have a lower aggregate cement ratio. In the wet mix process cement contents are usually in the range $350-450\,\mathrm{kg\,m^{-3}}$ with a water/cement ratio down to 0.4 if plasticizers are used. The mix has to be designed to be pumpable.

In the case of concrete repair all damaged, loose or hollow sounding concrete is broken away before applying sprayed concrete. The reinforcement is cleaned up if necessary, and the substrate is thoroughly dampened with fresh water. If it is necessary for additional reinforcement to be placed, there are certain design rules which should be followed to avoid blind areas and sand pockets and to permit the bars to be completely surrounded with well compacted concrete. The bars should be spaced at least 15 mm from the substrate if sand mixes are used and at least 50 mm from the backing concrete if coarse aggregate is included in the mix. The distance between parallel bars should be at least 65 mm. Lapping bars should not be touching but given a clear spacing of at least 50 mm. Where two layers of bars are to be used, the outer layer should be staggered and not placed directly over the inner layer. It should be remembered that any additional reinforcement in the repair will require cover appropriate to the exposure condition and the class of sprayed concrete.

The finished profile of sprayed concrete is often controlled by screeds. For flat surfaces the screeds may consist of rails of timber fixed to the substrate or reinforcement. In either case a gap should be left between the bottom of the rail and the substrate so that the rebound does not become trapped and can escape. Tightly stretched piano wires are sometimes used as screeds. In this case any excess concrete is removed using a knife of stainless steel, which is drawn over two adjacent wires. For less accurate work screeds are formed by shooting strips of sprayed concrete at appropriate intervals across the area to be covered.

When spraying columns it is usual to support screeding rails off two opposite sides and first spray the other sides. The screeds are then removed and the column is completed.

Correct spraying techniques are particularly important if blind areas and pockets of sand or rebound material are to be avoided. The nozzle should be held approximately 1 m from the surface being sprayed. A blowpipe operator directs a jet of air ahead of the nozzle to clear away rebound. On large areas it is usual to move the nozzle generally in a series of overlapping ellipse-shaped loops progressing first across the work area. At reinforcing bars and other similar obstructions, the stream should be angled in slightly from each side so that the material fills the gap behind. On vertical members spraying is started at the bottom to give support as the work proceeds upwards. The top edge of the work should be maintained at an angle of at least 45 ° so that most of the rebound falls clear.

Construction joints and daywork joints in sprayed concrete are treated differently to conventional work. The standard method is to form a shallow taper (say 4 : 1) at the edge of the work. The surface of the taper is brushed lightly before it sets to remove rebound and laitance. It may be necessary to clean up the surface with air or water if there is a delay in restarting. In any event, the joint should be thoroughly wetted before overspraying to complete the work. Joints can be formed in a similar way where work ends at a screed rail.

The sprayed concrete finish is usually left in the as-sprayed condition except that it may be brushed with a soft brush approximately 1 hour after placing. This is to reduce the occurrence of shrinkage cracks in the cement-rich outer layer. If trowelling is required, the timing is critical. At the stage just after the material has become hard, it can be cut with a trowel but the action tends to cause shallow cracking. This can be rectified by use of a steel or wooden float or by applying a further uniform thin flash coat.

As in any good quality concrete work, curing is essential if a high strength and durable material is to be obtained. On most sprayed concrete jobs the layers are thin and have a large surface area to volume ratio and efficient curing is even more important. The Concrete Society specification[6.2] requires that sprayed concrete should be protected from freezing or rapid drying out for a period of at least three days but notes that curing membranes must not be used if further layers of concrete or other bonded finishes are to be applied.

Quality control on sprayed concrete contracts often takes the form of test panels. These may be 750 mm square or bigger and 100 mm thick, and are fixed rigidly in position in the same orientation as the work surface. Test panels may be used before work commences for acceptance of materials, equipment and operatives, and also during the course of the work to check the strength being achieved. In either case 50 mm diameter cores can be cut for a visual assessment of compaction and presence of sand pockets, or 100 mm diameter cores can be cut for testing in compression. The test panel should be stored and cured alongside the actual work or under similar conditions. They are normally cored approximately 48 hours after spraying. Thereafter, the cores are stored under water or in conditions of high humidity according to the appropriate testing specification, BS 1881 Part 120,[6.5] for example. In the case of the wet mix process it is also possible to take samples of the concrete supplied to the gun and prepare cube or cylinder samples in the normal way, but this should be used to supplement rather than replace test panels.

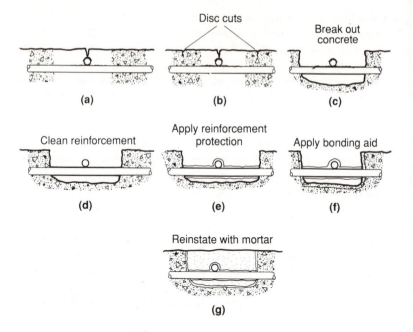

Fig. 6.3 Steps in patch repair process

Patch Repair

Patch repairs are discrete repairs carried out in small areas on a building or structure. They are generally less than half a square metre in area and are carried out using mortar applied by hand. The repair material may be cementitious, polymer-modified or a straight polymer mortar with an aggregate of fine sand or other filler. It is common practice in Europe to employ a patch repair system from one manufacturer. The repair system may include bonding aid, reinforcement primer, polymer-modified mortar, pore filler levelling mortar and protective coating(s). A typical sequence of operations is shown in Fig. 6.3.

Breaking Out

The first step in carrying out a patch repair is to cut a groove around the anticipated edges of the repair with a disc cutter to a depth of approximately 10 mm. This is to make sure that the repair does not taper out to a very shallow region (feather edge) when the main breaking out is undertaken. The shape of the repair is important and it should be delineated by straight lines. Abrupt changes in width or depth should be avoided if possible as these may result in cracking of the completed repair. Low length to width ratios are also to be preferred if practicable, but these may not be possible in all cases.

The main breaking out is usually undertaken by pneumatic or electric tools. These may cause shattering damage to the aggregate of the concrete which remains in-place unless it is carried out carefully. It is a noisy operation and may have to be restricted to certain times of the day when repairs are being carried out on occupied buildings or in residential areas. Cutting by high pressure water jet is a quieter alternative which causes less damage to the substrate. However, the equipment is not widely available and this method is more expensive than mechanical breaking. Where repair is being carried out on a structure with corroded reinforcement, the condition of the exposed reinforcement is checked at this stage. If corrosion is present on the bars at the edge of the repair area, additional concrete is broken out. Breaking out continues until at least 50 mm of reinforcement free from active corrosion is encountered in all directions. The edges of any additional breaking out are also cut square with a disc cutter. Concrete is usually cut out beyond the reinforcement to give a clear gap of approximately 20 mm. This is done so that the backs of the bars can be properly cleaned and also so that the completed repair is effectively locked in.

Reinforcement

Exposed bars are cleaned at this stage by grit blasting. Other methods such as mechanical wire brushing are not as effective in removing corrosion or result in an inappropriate surface profile. Grit blasting is also one of the best methods of treating the backs of bars and the difficult areas where bars cross or lap. Various manufacturers of repair systems specify different qualities of surface preparation before application of the primer, but the best possible standard should be achieved in all cases. Research has indicated[6.6] that corrosion in the repair is less likely to be initiated if all rust has been removed from the bars.

Where the reinforcement is deeply pitted, and particularly where it has been exposed to high concentrations of chlorides, it may be extremely difficult to remove all corrosion with a single grit blasting treatment. It may be found that corrosion starts again very quickly in the base area of the pits. In these circumstances grit blasting followed by washing with fresh water followed by further blasting has proved to be effective. In severe cases it may be simpler to cut out the affected length and lap in a new bar.

Supplementing or replacing the existing reinforcement may also be necessary if a significant percentage of the area has been lost through corrosion. A judgement on the need for additional steel can be made on the basis of an analysis of the particular section but it should be kept in mind that, unless the member has been propped to relieve the effects of dead load, the supplementary reinforcement can only assist in carrying the live loads. Many patch repairs are too small to permit the full laps to be achieved. In these cases it may be possible to employ welding or to use mechanical couplers or splices. Not all reinforcing steels are weldable. Guidance on this topic is given in a British Steel Corporation publication.[6.7] In many repair situations there may not be enough cover to allow mechanical couplers to be used. In these circumstances it may be necessary to break out additional concrete locally so that the full lap length can be achieved. All replacement steel should be grit blasted to the same high standard as the existing reinforcement within the repair.

In cases where one of the causes of deterioration was lack of adequate concrete cover to the reinforcement, the bars should be bent back or otherwise moved to achieve sufficient cover in the repair. If this is not possible an alternative is to build out the repair so that it finishes proud of the surrounding surface but this is not always an aesthetically acceptable solution.

Once the reinforcement has been prepared by grit blasting the first coat of reinforcement primer should be applied within say 3 hours. In a marine environment or under humid conditions, a shorter time lapse will be appropriate. The primer should totally encapsulate the exposed reinforcement paying particular attention to locations where bars cross.

Reinstatement

The next step is to prime the surface of the concrete within the repair area. The actual priming treatment will depend on the nature of the repair mortar.

For cementitious repair mortars the substrate is thoroughly soaked with clean water before applying a bonding aid. The bonding aid may consist of a polymer emulsion or a polymer cement slurry. The function of the presoaking and also the bonding aid is to reduce the suction of the substrate and hence the amount of water lost from the repair mortar. In hot climates the water applied to the surface is taken up very rapidly by the concrete and also evaporates very quickly. A prolonged period of soaking, possibly overnight, may be appropriate in these circumstances. When the bonding aid is applied the surface should be damp but free of standing water.

In polymer repair mortars a bonding aid of pure polymer is worked into the concrete surface without soaking with water. In most cases the concrete surface should be dry at the time of application but some polymer bonding aids tolerant of damp conditions are available.

The timing of the application of the first layer of repair mortar is also dependent on material type. For most systems which incorporate cementitious mortars, the mortar must be applied while the bonding aid is wet. It has been found[6.8] that if the bonding aid is allowed to dry before the mortar is applied, there may be a critical reduction in the bond between repair and substrate. For polymer mortars the bonding aid should be allowed to reach a tacky condition before application of mortar commences. The time taken to reach this condition will vary according to the polymer type, the ambient temperature and the temperature of the substrate.

Both polymer and cementitious mortar patch repairs are built up in layers. The materials have to be hand-placed using a technique which results in good compaction and density. Normal techniques employed in plastering operations are designed to produce an even spread of material and are not appropriate. The mortar has to be carefully worked behind and around reinforcing bars to give good protection. The thickness of individual layers in a repair will depend on the material itself and the orientation of the surface being repaired. Layer thicknesses of 25–50 mm are typical for vertical repairs and 20–30 mm for overhead repairs with normal weight mortars. Mortars with lightweight aggregates have been developed so that greater layer thickness can be achieved in overhead work.

The surface of intermediate layers is usually left in a rough condition to assist in producing a good bond to the next layer. The following layer is placed when the previous layer has stiffened sufficiently to carry the applied weight but before final set. If the following layer is delayed, the surface of the previous layer is scratched with a trowel, dampened with water and a bonding aid is applied before the work continues.

The final layer should be at least 10 mm thick. The finishing on flat areas can be achieved by steel or wooden float to match the surrounding concrete. With good technique it is possible to obtain a reasonable match with ribbed and other decorative finishes. Exposed aggregate finishes have been successfully matched by pressing stones individually into the final layer whilst it is still in the plastic state.

Effective curing is important. For polymer mortars the requirement is to protect from cold, heat or rain. For polymer-modified cementitious mortar the requirements are similar but with the additional requirement that rapid drying out should be prevented. This can be achieved by efficient spray-on curing membranes if no subsequent decorative or other treatments are to be applied. An alternative is to use sheets of polythene taped securely to the structure around the repair.

Materials

Patch repairs have been used for many years. Originally sand and cement were used and the constituents were batched at site using a minimum of water to produce 'dry pack'. At a later date the practice was modified to include the use of a polymer latex such as styrene–butadeine as part of the gauging water. Modern procedure is for all of the repair materials to be provided in a packaged form as a system from one manufacturer. As far as mortars are concerned, preblended dry sand and cement may be supplied in one package, and the latex and the appropriate quantity of water in another. The ingredients simply have to be mixed together thoroughly at site to produce the repair mortar. Another alternative is to use redispersible polymer powders which are manufactured by spray drying. These are blended at the factory with the appropriate quantity of dry graded sand, cement and other admixtures in powder form. In this case only water has to be added at site. This change of practice towards factory blended materials has led to improvements in quality control and a reduction in the number of unsatisfactory repairs.

The most commonly used polymers in cementitious repair mortars are styrene–butadiene (sometimes known as SBR) and acrylics. Two distinct structural forms develop at the microscopic level as the mortars set and gain strength. There is the matrix of cement hydrates as would be expected, but this is interwoven with a network of polymer strands. The addition of polymers to repair mortars improves their properties both before, during and after setting:[6.9]

1. When the mortar is in the plastic state they have a plasticizing effect. This reduces the amount of water that has to be added to produce a workable mortar and hence reduces the amount of long term shrinkage.
2. There is an improvement in the bond between repair mortar and substrate.
3. The polymer tends to reduce the loss of water from the repair and therefore acts to some degree as an internal curing aid.
4. The polymer strands reduce the degree of microcracking as they bind the structure together. This results in
   (a) increase in flexural and tensile strength;
   (b) reduction in permeability to moisture and carbon dioxide.

The bonding aid used on the original concrete surface of the repair is usually a slurry of polymer and cement which reduces loss of moisture from the repair and also improves bond.

Several different forms of reinforcement protection are used in repair systems. They may act purely as barriers to protect the steel from moisture and other potentially damaging substances or they may act by giving chemical or electrochemical protection. Barrier coatings are usually of epoxy and they require a high standard of surface preparation on the reinforcement. In some cases two coats are applied and a blinding of quartz sand is used with the second coat to improve bond. Cement polymer slurries give additional protection as they lead to highly alkaline conditions around the reinforcement which has a passivating effect. Some epoxy reinforcement protection materials also contain finely ground cement clinker in an attempt to provide an alkaline environment. The clinker also acts as a moisture scavenger. Powdered zinc is used in some epoxy primers to introduce an element of electrochemical protection akin to galvanizing.[6.10]

Testing

Few tests have been developed specifically for the control of patch repair works but some of the methods used for concrete can be adapted. As an example cubes of repair mortar can be made at site and tested for compressive strength. The cubes are usually much smaller than those used for concrete testing. A cube of size 40−75 mm is normally employed because of the smaller amounts of material used and also so that the cube dimensions are of the same order as the depth of most repairs, and curing conditions in the field and for control specimens are similar. Unlike concrete test cubes, mortar cubes are not maintained in a saturated condition but allowed to dry out naturally. Many of the polymers used in cementitious mortars have to lose water in order to polymerize and this would not occur if the test cubes were retained under water.

It is also useful to core into some repairs so that the standard of compaction can be judged on a visual basis or by testing the density. The taking of cores can be combined with pull-off testing to check the bond between repair and substrate, and also the bond between the various layers which make up the complete repair.

When judging the results of tests for compressive strength, pull-off strength and density it is not appropriate to compare directly with the results quoted on manufacturer's technical information data sheets. It is probable that manufacturer's data relates to specimens manufactured and cured under ideal laboratory conditions. Just as concrete in structures does not necessarily achieve the strength indicated by the control cubes neither can mortars in-place in repairs be expected to achieve the same properties as laboratory produced specimens.

Bulky Repairs

In repairs of large volume it may not be appropriate or economic to use hand-applied mortars or sprayed concrete. In this situation concrete in shuttering is often used. Although some very large volume repairs with good access have been carried out in this way, in many cases there are only very limited openings in the shutters and access for vibration may be extremely restricted or non-existent. Flowing concrete which requires little compaction has been developed to overcome this difficulty.

The preparation of the repair area before concreting is similar to that for patch repair but on a bigger scale. All loose and damaged concrete is broken out, the edges are cut back square so that feather-edges are avoided and the reinforcement is thoroughly cleaned. Bonding aids are sometimes used, particularly on thin pours such as some repairs to slabs. Where bonding aids are used on shuttered repairs the shutters have to be carefully designed so that they can be positioned quickly before the bonding aid dries.

Concrete has to be introduced carefully into the mould so as to avoid entrapment of air. Pumping is usually employed but concrete may be placed conventionally through a series of 'letter box' shutters which are closed successively as the concrete attains their level. This technique is illustrated in Fig. 6.4. Where pumping is used it is best to introduce

Fig. 6.4 Letter-box shutter for conventionally placed concrete

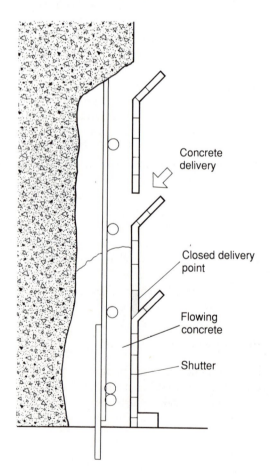

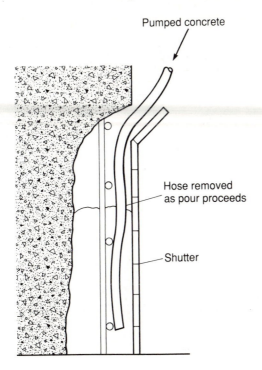

Pumped concrete

Hose removed
as pour proceeds

Shutter

Fig. 6.5 Letter-box
shutter for
pumped concrete

the delivery hose at a low position in the pour and allow the air to be displaced vertically as the concrete fills the mould. This can be achieved fairly easily on the sides of members as shown in Fig. 6.5, but it is not so easy for repairs to soffits. In this case it may be necessary to use a second pipe as a vent at the highest position in the pour (Fig. 6.6). It may be necessary to break out additional concrete so that the upper surface of the pour is well shaped and so that dead areas where air is likely to be trapped are avoided (Fig. 6.7).

In some large volume repairs a technique of grouting preplaced aggregate is used, as shown in Fig. 6.8. If necessary formwork is placed in position as for conventional concreting and the pour is filled with single

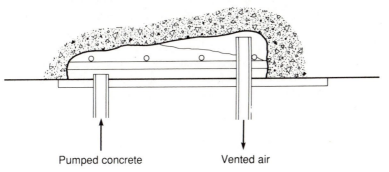

Pumped concrete   Vented air

Fig. 6.6 Pumped
concrete repair on
soffit

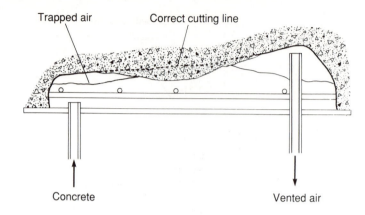

Fig. 6.7 Consequences of incorrect break-out on soffit repair

Trapped air   Correct cutting line

Concrete   Vented air

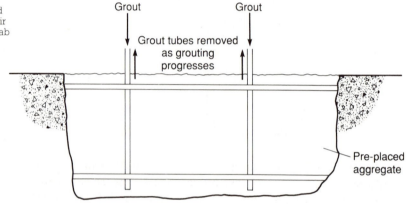

Fig. 6.8 Grouted aggregate repair of deep floor slab

Grout   Grout

Grout tubes removed as grouting progresses

Pre-placed aggregate

size coarse aggregate. Grout is pumped in through a pipe into the bottom of the pour. If the pipe passes through the pour it is gradually withdrawn as the grout level rises.

Crack Filling and Resin Injection

Resin injection is used to fill cracks and to bond together the concrete surfaces. It is essential to determine the cause of cracking and choose a resin with suitable properties or the procedure may not prove to be an effective long term solution. For instance, if shrinkage cracks are filled with a rigid resin and shrinkage continues, it is probable that new cracks will form in the concrete alongside the old as concrete is much weaker in tension than most injection resins. Resins with low modulus are available and these allow continuing movement at the cracks. Injection is not usually considered to be an acceptable method of treating cracks such as those due to reinforcement corrosion where an underlying continuing process has been responsible for their generation.

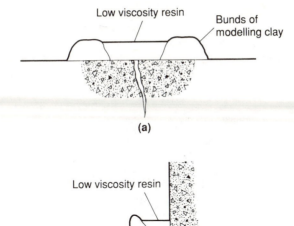

Fig. 6.9 Methods of filling cracks by gravity.
(a) Horizontal surface;
(b) vertical surface

Cracks are often filled to prevent moisture, atmospheric gases and other potentially deleterious substances from finding a relatively easy pathway to the reinforcement. The process does not necessarily have to involve complex equipment. For example, plastic shrinkage cracks on the top surface of a slab are often relatively wide and can be filled by ponding using a low viscosity resin under gravity. The cracks are first blown out using dry oil-free compressed air. Reservoirs are formed on the surface around the cracks using modelling clay and the resin is poured in, as shown in Fig. 6.9(a). It is possible to use a similar procedure for cracks in vertical surfaces. In this case 'birds nest' reservoirs are formed over the crack as shown in Fig. 6.9(b).

More usually it is necessary to apply a positive or negative pressure to the resin to assist its flow into the crack. A positive pressure system is illustrated in Fig. 6.10(a). The surface of the crack is first cleaned out and then sealed using a polyester putty or other suitable material. Injection ports are cemented to the surface or short copper pipes are drilled and cemented in to intercept the crack (Fig. 6.10(b)). At this stage the sealing of the system can be checked by applying a low air pressure to one of the ports with the others sealed. If the pressure cannot be maintained, leaks can be detected by smearing soapy water on the surface.

Fig. 6.10 Resin injection under positive pressure

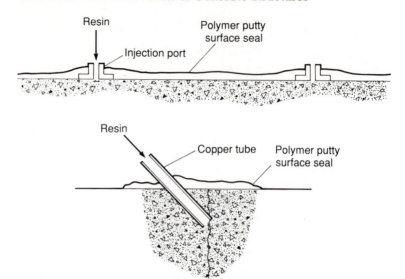

When resin is injected into a crack it spreads out in a semi-circular pattern with the injection port at its centre as shown in Fig. 6.11. Injection ports are therefore spaced at slightly less than the member thickness or crack depth in the case where injection is from one side. If injection can be carried out from both sides the injection ports are placed at a spacing of slightly less than half of the member thickness.

Fig. 6.11 Spread of resin from injection port

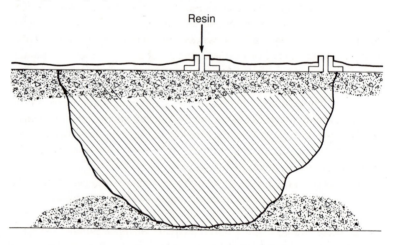

Materials

Resins usually consist of two components, one is the active ingredient and the other the catalyst or hardener. In some cases where the cracks are wider or where larger voids are to be treated an inert filler is also included. This has the effect of reducing shrinkage as the resin cures and also limiting temperature rise as the exothermic reaction takes place. The components have to be thoroughly mixed together in carefully controlled proportions if a final product with the desired properties is to be produced. Two methods of mixing and injection are in common use. In the first the components are combined in a separate mixer and then transferred to the application equipment. The positive pressure for injection may be supplied by use of an electrical pump, compressed air or a hand piston pump similar to a grease gun or a cartridge gun used in sealant application. In the second method the components are pumped separately in carefully controlled proportions to a mixing head where mixing takes place automatically as the two streams flow together. The latter method is suitable for large volume work in the situation where continuity of the process over a long period is assured. If there are any prolonged stoppages in the process there is the risk that mixed resin will start to harden in the mixing head. The former method is more suitable for smaller jobs where the work is intermittent.

The majority of injection work is carried out with free flowing resins of low viscosity. However, other materials are used in special circumstances. Thixotropic formulations are used where cracks extend through the thickness of a member, but access can only be gained from one side. These materials flow when pressure is applied but remain in position when pressure is removed. This property reduces the likelihood of excessive loss of material on the blind side. Specially formulated thixotropic gels are sometimes injected into cracks where there are water seepage problems. The gels are designed to expand when they come into contact with water and block off the cracks.

Method

Injection starts by introducing resin under pressure at the first port with the other ports open. In the case of vertical cracks it is usual to start at the lowest port. Injection continues until resin appears at the second port; the first port is sealed off and injection starts at the second port and the process is repeated. While injection is proceeding the structure is carefully monitored for signs of resin leaks from interconnected cracks or joints.

A variation on the above procedure using negative pressure has been developed in the United Kingdom in what is understood to be a patented process. As resin is supplied under positive pressure to one port, vacuum is applied to one or more of the neighbouring ports drawing the resin towards them. This process can give much more control over the spread of resin within a crack system.

en injection is complete the injection ports are removed and the
ce sealer is scraped off. It may be necessary to apply gentle heat
sist in this process and also to use a fine grinding wheel.

Testing

ere are a number of control and testing systems which can be used
during resin injection work. As each consignment of resin is received
at site its setting time can be checked by mixing together a particular
quantity of material in the appropriate proportions and noting the
hardening time. Clearly this is not a particularly refined test, but if it
is repeated under approximately similar conditions it should pick up any
gross anomalies in formulation. The quantities of resin injected into a
crack system and the resin takes at each port can be recorded. This enables
areas of difficulty to be located where cracks may not have been
completely filled or where hidden voids are present. Finally, cores which
include the crack can be cut and examined. Core diameter will depend
on crack geometry but 25 mm diameter cores have been used successfully
in some cases. The locations for coring can be chosen on the basis of
the records of resin take. Cores can be examined by eye for completeness
and depth of resin penetration if the cracks are wide. For finer cracks
it will be necessary to produce polished sections and examine them under
the microscope.

Bonding of External Reinforcement

Steel plates bonded to the concrete surface have been used to provide
additional strength or stiffness to structures. The technique has been used
in the case where change of use has required an office floor to be uprated
and also where strengthening of a bridge has been required to facilitate
the passage of an exceptional load. Some of the first applications were
carried out in the 1960s[6.11] and the technique has become more
widespread with the development of epoxy adhesives of great strength.
There are few reports as yet of durability studies on strengthened
structures.

The normal application of the technique is to increase flexural capacity
by bonding plates to the tension side. The concrete surface must first
be carefully examined to make sure it is free from any defects likely
to have a detrimental effect on the transfer of stress between concrete
and steel. Any weak friable surface or coating must be removed. The
bonding surface of the plate is cleaned to a high standard using grit
blasting. The adhesive is applied within a short period and the plate is
pressed against the coated surface and supported in position until the
adhesive has cured. Adhesives having a paste-like consistency are usually

employed and typically the adhesive layer thickness is of the order of 1−2 mm.

When plates are used on the soffits of beams or slabs there is a danger that they could become detached if a fire occurs. In these circumstances it is prudent to provide some secondary positive fixing system between the steel plate and the concrete such as bolts into mechanical anchors drilled into the concrete.

References

6.1   Concrete Society 1980 *Code of Practice for Sprayed Concrete* The Concrete Society, London

6.2   Concrete Society 1979 *Specification for Sprayed Concrete* The Concrete Society, London

6.3   Concrete Society 1981 *Guidance Notes on the Method of Measurement for Sprayed Concrete* The Concrete Society, London

6.4   Ryan T F 1973 *Gunite — A Handbook for Engineers* The Cement and Concrete Association, Wexham, United Kingdom

6.5   British Standards Institution 1983 *Testing Concrete. Method for Determination of the Compressive Strength of Concrete Cores* BS 1881: Part 120, The British Standards Institution, London

6.6   John D G 1981 Novel electrochemical techniques for investigating steel/ concrete systems. Proceedings of a One Day Conference on Failure and Repair of Corroded Reinforced Concrete Structures IBC, London, pp 59−86

6.7   Reinforcement Steel Services 1976 *The Welding of Reinforcing Steel — A Guide to Design and Procedures* British Steel Corporation, Middlesbrough

6.8   Dixon J F 1983 Use of bond coats in concrete repair. *Concrete* **17 (8)**: pp 34−5

6.9   Shaw J D N 1987 Polymers for concrete repair. In Allen R T, Edwards S C (eds) *The Repair of Concrete Structures* Blackie, Glasgow and London, pp 37−52

6.10   McCurrich L H, Cheriton L W, Little D R 1985. Repair systems for preventing further corrosion in damaged reinforced concrete. *Proceedings of the First International Conference on Deterioration and Repair of Reinforced Concrete in the Arabian Gulf* Bahrain Society of Engineers, Bahrain, pp 151−68

6.11   Jones R, Swamy R N, Salman F A R 1985 Structural implications of repairing by epoxy bonded steel plates. *Proceedings of The Second International Conference on Structural Faults and Repair* Engineering Technics Press, Edinburgh, pp 75−80

7 Protection

Introduction

In most cases the processes which were the cause of deterioration and which had led to the original need to undertake renovation, will continue to act after repairs have been carried out. Although the chosen repair methods should provide adequate protection at the individual repair sites the remainder of the surface will remain directly exposed to the environment and there is the possibility of further deterioration within a short time period. This being the case and taking into account the future requirements of the structure, some form of additional general protection is often included in renovation contracts.

Additional protection of concrete structures after construction is usually either in the form of a surface treatment (coatings, impregnations) or the electrical system known as cathodic protection. In recent years a process of removing chlorides and restoring alkaline conditions using electrolytic techniques has been developed.

Coatings are frequently used in the case where carbonation has resulted in corrosion of reinforcement. The areas of cracking and spalling are broken out and repaired but in adjacent areas the carbonation front may be close to the reinforcement. If carbonation were allowed to continue unchecked there could be further cracking and spalling within a short period. 'Anticarbonation' coatings are designed to reduce the penetration of carbon dioxide, oxygen and moisture, and thereby extend the time before further repairs become necessary.

The most usual application of cathodic protection to reinforced concrete structures is in the case of chloride contamination. If only cracked and spalled concrete is replaced in such structures, large areas of reinforcement will remain in contact with concrete containing significant concentrations of chlorides. It has been suggested[7.1] that breaking out and repairing in this way may in fact increase the potential for reinforcement corrosion in adjacent areas. Sometimes attempts are made to break out all chloride-contaminated concrete but this can prove to be expensive and it is difficult to identify the limits of chloride contamination. In these situations it may be appropriate to consider undertaking localized repair of damaged areas and the application of a cathodic protection system.

It should be remembered, and pointed out to the structures owner, that the protection itself may impose its own maintenance liability. For example surface coatings generally have a life in the range of 5−15 years and have to be renewed or overcoated. Most applications of cathodic protection to reinforced concrete are relatively recent and little is known about the practical life expectancy of anodes, overlays or control systems.

Surface Treatments

Surface treatments have been used on concrete masonry and render for decorative purposes for many years, but their use to provide protection has been developed mainly since about 1970.

Many deterioration processes result from penetration of the concrete by gases or salts in solution. Surface treatments are designed to reduce the passage of potentially deleterious substances and hence slow down the rates of deterioration.

A wide variety of materials has been used in attempts to provide protection but products can generally be classified in one of three main groups according to the means by which they provide protective action:[7.2]

(a) coatings/renderings;
(b) pore liners/pore blockers; and
(c) sealers.

This classification system is useful but there may be some products which cannot be clearly classified because they are designed to provide protection in two or more ways. As examples, many coatings penetrate and seal the surface to a certain extent and many sealers form a film and coat the surface.

The essential differences between the groups of products are illustrated in Fig. 7.1. Gases and fluids enter concrete mainly through the pore system. Coatings and renderings protect by providing a layer of material on the surface which spans across the pores. Pore liners or pore blockers penetrate into the pores and either provide physical blocks or make the surfaces water repellant. Sealers are part way between coatings and renderings. They partly penetrate the pore system but also form a thin film on the surface.

Fig. 7.1
Differences
between the
actions of coatings,
pore blockers and
pore liners

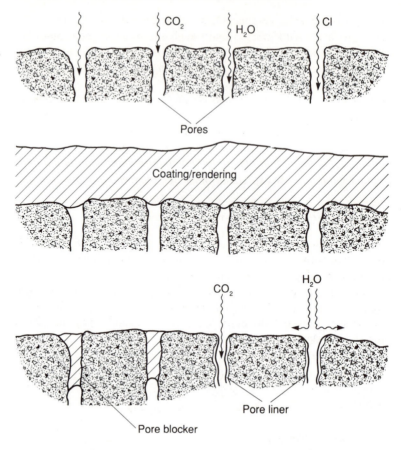

Coatings/Renderings

There is no clear line of distinction between coatings and renderings as far as material properties or functions are concerned. Renderings are generally bound by hydraulic cement and they are defined in this way in a British Standard[7.3] relating to the building trades.

However, polymer-bound renderings are also available and are frequently used to protect concrete in chemical works and sewage treatment plants in hot climates. The fact that the main binding material in a surface protection product is a hydraulic cement does not necessarily mean that it should be classified as a rendering as cement-based coatings are widely available and are frequently used on masonry.

The distinction in the context of concrete rehabilitation and protection is mainly one of thickness and method of application. The CIRIA report cited above[7.2] suggests that materials with an applied total thickness of less than 5.0 mm should be described as coatings, whereas materials with

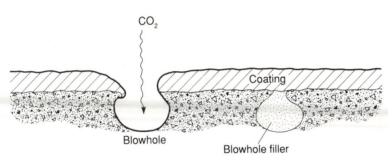

Fig. 7.2 Effect of
blowholes on
integrity of coating

greater thickness should be described as renderings. A further subdivision
into thin and thick coatings is suggested with thick coatings having a
thickness greater than 1.0 mm. The method of application is also a guide
as to whether a material should be considered as a coating or a rendering
but it is not infallible. Coatings are generally applied by methods suitable
for paint treatments such as brush, roller or spray whereas renderings
are often applied by plastering techniques such as trowelling.

Coatings and renderings provide protection by creating an external
physical barrier which slows down the penetration of liquids and gases.
In the case of coatings the applied material forms a thin coherent film.
The film-forming material is often a polymer usually with a filler to give
bulk and thickness and a pigment. Renderings usually contain a much
higher proportion of inert mineral or other fillers. If based on hydraulic
cement they may also contain polymers which act as plasticizers and
also reduce the permeability of the hardened rendering. Some renderings
may also include fibres with the aim of improving cohesion in the plastic
state and also reducing cracking and improving impact properties in the
hardened state.

Some terms and materials are in common use in the concrete protection
industry which have different meanings or no equivalent in the
plastering/rendering trades. Pore fillers, also known as pore stoppers,
are used to fill blowholes (bugholes) or other small surface defects before
applying coatings. The terminology is unfortunate as these materials are
designed to deal with features with a diameter in the range $1-5$ mm and
not the capillary pores described below in the context of pore liners and
blockers. Pore fillers are used because it has been found that untreated
blowholes can cause pinholes in coating films as shown in Fig. 7.2.
Pinholes allow localized carbonation to continue and could result in
intense localized corrosion of underlying reinforcement.

Fairing coats or levelling coats are used on rough concrete surfaces
to produce a smoother finish suitable to receive a thin coating. If thin
coatings were applied directly to rough surfaces the film thickness
achieved could vary considerably with thick areas in depressions and
thin areas on peaks with the possibility of some discontinuities, as shown

Fig. 7.3 Effect of
surface roughness
on thickness and
integrity of coating

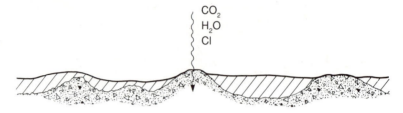

in Fig. 7.3. This could lead to localized corrosion of reinforcement as in the case of unfilled blowholes described above.

Pore Liners/Pore Blockers

Pore liners and pore blockers are applied as liquids of low viscosity which are capable of penetrating the minute capillary pores in the concrete surface. Once in position they solidify by solvent evaporation, crystallization or polymerization to produce a physical plug or react with materials present in concrete to produce hydrophobic compounds which cover the surface and line the pores.

Hydrophobic compounds change the wetting characteristics of the surface. Water under low pressure tends to form droplets on the surface and the ability to enter the pores is reduced. The ingress of chlorides is therefore restricted as they usually enter concrete in solution. However, water under pressure is able to penetrate the lined pores and these hydrophobic materials are not generally recommended even in the case where ponding is likely to occur.

Pore liners and pore blockers have advantages over coatings for some applications. As they do not form an exterior film and are colourless, they result in little change in the appearance of the concrete. The materials are capable of penetrating well into the surface and if this is the case, they are not much affected by ultraviolet radiation and other weathering agents. Their protective properties can therefore be maintained over a long period.

In the case of pore lining materials, the pores are left open to the passage of water vapour; the concrete is able to breathe. This should result, in the long term, in the drying out of the surface layer of concrete as the net intake of moisture is reduced. This could result in a net reduction in corrosion activity as discussed later.

The passage of gases from the atmosphere, oxygen and carbon dioxide, is also permitted by pore liners and may in fact be increased when compared with untreated concrete. This is because the pores of untreated concrete become full of water when the concrete is in a saturated condition, for instance, after rainfall, and the gases cannot penetrate. There has been concern that this could result in increased rates of carbonation. However, rates of carbonation are very much reduced in

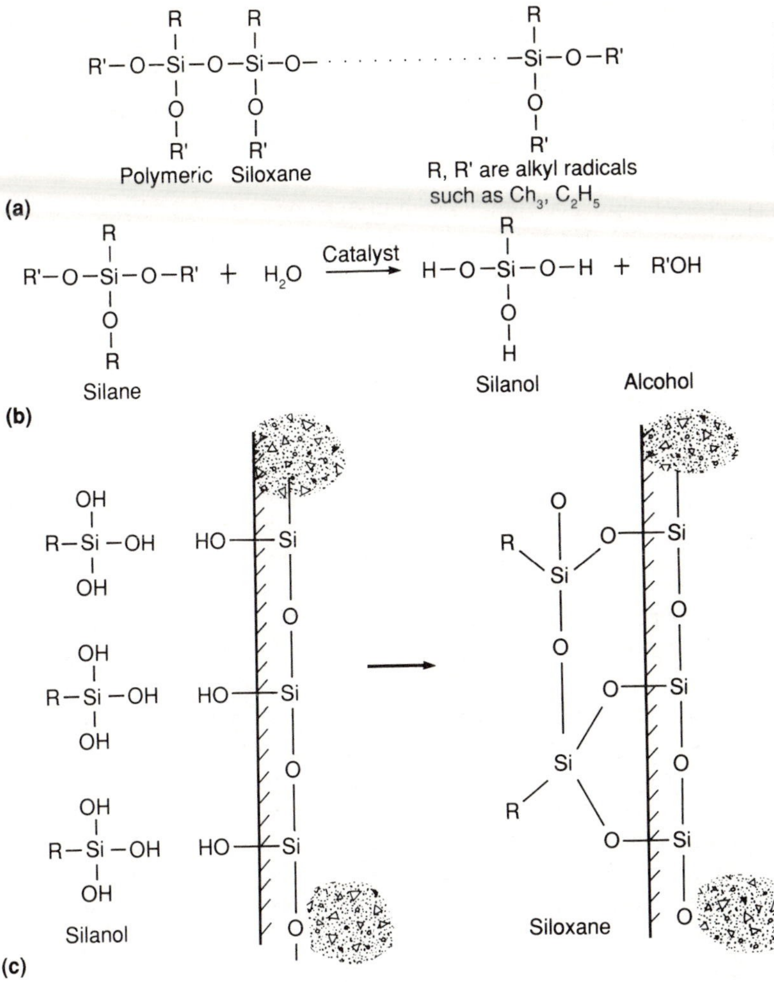

Fig. 7.4 Chemical
reaction in
producing
hydrophobic
surface on
concrete using
silane

(a)

(b)

(c)

dry concrete and if this is taken into account the overall effect may not
be very great.

One interesting group of compounds used as pore liners and pore
blockers are those based on silicon and known as silicones. The
differences in the chemical structures of the compounds and the effects
of these differences on their use as water repellents has been described
in papers by Roth[7.4] and Rodder.[7.5] The main points are summarized
below.

Silicon, like carbon, is capable of producing long chain compounds.
In the case of silicon, the atoms alternate with oxygen atoms as the links
in the main chain and alkyl radicals are also present, as shown in Fig.
7.4(a). These long chain compounds are known as siloxanes. Under
suitable conditions they can be cross-linked to produce silicone resins

which are known as polysiloxanes. The properties of the silicone resins will vary according to the size of the attached alkyl radicals but many have waterproofing properties. The silicone resins can be dissolved in organic solvents. When applied in solution to concrete surfaces they penetrate to shallow depths and produce a water-repellent surface when the solvent evaporates.

The shallow penetration depth of silicone resin solutions is a result of the size of the resin molecules. Better penetration depths are achieved by applying lower molecular weight reactive silanes to the surface. In the presence of water, the silanes produce an intermediate product called silanol and alcohol as shown in Fig. 7.4(b). The silane polymerizes and undergoes cross-linking in position on the surface. It also takes part in reactions with silicate molecules in the hydrated concrete and so is chemically bound into the surface (Fig. 7.4 (c)).

A potential disadvantage in the use of silanes is that they require the presence of moisture to undergo the conversion reaction to silanol and they are also relatively volatile. There is therefore the possibility that the applied material may evaporate before the reaction has been completed, if the substrate conditions are not favourable. Another possibility, which overcomes these potential problems, is the use of short chain polysiloxanes. These clearly have a larger molecular size than silanes but they are less volatile. They can therefore remain in-place for much longer until conditions are right for further polymerization and cross linking to give the surface water-repellent properties.

As mentioned previously, treatment with hydrophobic agents such as silane should permit the surface of the concrete to dry out over a period of time. If the drying out progresses to the level of the reinforcement it removes one of the prerequisites for the occurrence of corrosion. It is theroretically possible that removal of water could stifle corrosion even in the situation where high concentrations of chloride are present. A small research programme[7.6] has been carried out in the United Kingdom to determine whether the theoretical possibility is borne out in practice.

The investigation has been undertaken on two crossbeams on the substructure of Wolvercote Viaduct in Oxfordshire. The crossbeams often become saturated with chloride-laden water because of rundown and leakage from the deck above, and it was found that there were significant concentrations of chloride at the level of the reinforcement. In areas of low cover, corrosion of the reinforcement and spalling of concrete had occurred. Half-cell potential mapping revealed widely varying potentials with some peaks having values reaching $-540\,mV$ (silver/silver chloride). Breaking out at some of these locations revealed the existence of pitting corrosion.

The cross-heads were repaired by breaking out cracked and spalled areas and reinstating with concrete or mortar. Attention was also paid to improving the drainage of the cross-heads. An acrylic waterproof coating was applied to the top surfaces of the beams and silane was applied to the sides and soffits.

Automatic data-logging facilities were installed as part of the remedial works contract. These have permitted monitoring of potential and resistivity using built-in probes in addition to the ambient conditions of relative humidity and temperature. Initial results after one year's monitoring are encouraging in that they have shown a general decrease in the electrode potentials and a general increase in the measured resistivity, which appears to be indicative of drying out of the concrete.

Another group of silicon compounds which have been widely used to treat concrete surfaces for many years are surface hardeners containing silicates or silicofluorides. These are applied as solutions in water and react with lime present in the matrix to form insoluble calcium silicate which crystallizes in the concrete pores. The crystals partially block the pores and reduce the permeability. The free lime in a concrete surface may carbonate fairly rapidly after construction and materials of this type may have a much reduced effect in these circumstances.

Sealers

Sealers are intermediate between coatings and pore blockers. They have penetrating properties, which may be achieved by solventing, and also form relatively thin surface films. Materials of this type may be used to bind together and strengthen weak friable surfaces. Also because of their penetrating properties and the resulting firm adhesion to the substrate, they are used to prime surfaces before the application of coatings.

Application of Surface Treatments

A wide variety of families of chemical products are used as principal components in surface treatment products. Acrylics, epoxies, polyesters, chlorinated rubbers and polyurethanes are all used in coatings either in polymer dispersions, in solvents or in reactive two-component form. They can be applied by brush, roller or spray.

Careful attention has to be paid to surface preparation and conditions during application if a successful and durable result is to be achieved. The particular requirements will vary according to the individual product and there is a need for a precise specification from the manufacturer. This should give clear information on the strength/condition of the concrete surface and its cleanliness, the moisture condition of the concrete and the ambient temperature and relative humidity required for the application of the product.

Concrete surfaces to receive protective treatment are often prepared and cleaned by wet or dry grit blasting. This is capable of removing old surface treatments and surface contaminants and also has the advantage that it provides an appropriate surface texture to receive coatings. It opens up blowholes and the blowholes will have to be filled, as mentioned previously, to reduce the occurrence of pinholes in the coating.

After grit blasting, dust may remain on the surface. This can be removed by blowing off with air from a compressor as long as steps have been taken to remove oil from the stream. Some coating systems are moisture sensitive and it may be necessary to allow the surface to dry out if wet blasting has been used. On the other hand, cementitious renderings are best applied to a damp surface as otherwise suction from the underlying concrete may remove sufficient water from the rendering to affect the efficiency of its hydration. Useful information on the preparation of surfaces for coatings and simple practical acceptance tests have been given by Gaul.[7.7]

Surface penetrants such as silanes and siloxanes are applied by flooding horizontal surfaces or by low pressure spray on vertical and overhead surfaces. As they must penetrate to be effective, it is essential that the surfaces are not saturated when the material is applied. Similarly the materials take time to react and should not be applied in windy or warm conditions when evaporation is likely to take place before the reaction has progressed sufficiently. Guidance on application conditions has been given by the United Kingdom Department of Transport.[7.8] They suggest that the surface should be kept dry for a minimum of 24 hours prior to application and also be protected from rain and traffic spray for at least 6 hours after application. They also suggest that application should not proceed when the shade temperature is less then 5 °C nor when the concrete surface is at a temperature greater than 25 °C.

Very few methods are readily available to control surface treatment applications at site. Control of film thickness is important because if the application is too thin it will not give the desired degree of protection. If the coating is too thick it may give rise to curing and adhesion problems in addition to being uneconomic. None of the methods usually used for measuring wet film thicknesses on metal substrates[7.9] is appropriate for the rough surfaces normally found on concrete. One alternative is to take cores and measure dry film thickness on cut sections. Another, but less satisfactory alternative, is to attach a thin sheet of metal or board to the surface before coating. The sheet is coated along with the surface and the wet or dry film thickness can be measured by any appropriate means. As a very broad check on application rates, the quantity of coating used each day or half-day can be recorded and related to the area of surface that has been coated.

Comparative laboratory studies on protective properties of surface treatments have been carried out in Europe and the United States. The work in the United States[7.10] was carried out for the Transportation Research Board and concentrated on the protection of bridges against penetration of chlorides. Initially 21 surface treatments including epoxies, methacrylates, urethanes, butadienes and a silane were subjected to a preliminary screening test. The treatments were applied to concrete specimens and exposed in a 15 per cent salt water solution. The water absorption was monitored by weight gain over a 21-day period. Water vapour transmission properties were also measured during a subsequent air drying period. Chloride contents of the concrete in the specimens were determined after the soaking period. It was found that there was good correlation between chloride content and weight gain.

On the basis of this initial screening programme, five products with low water absorption, low chloride ion uptake and good water vapour transmission characteristics were chosen for further testing. The five materials chosen were an epoxy, a methyl methacrylate, a moisture-cured urethane, a silane and a polyisobutyl methacrylate. These materials were subjected to further testing to determine the effects of the moisture condition of the substrate, coverage rate and different environmental conditions on the ability to protect against ingress of chloride solutions. The report on the work concluded that the epoxy, the methyl methacrylate and the silane were capable of providing added protection to concrete bridge surfaces to reduce intrusion of salt laden water. However, the report also recommended that before using a formulation based on these materials, evaluation tests similar to those described should be carried out on the particular products.

Research in Europe has tended to concentrate on the protection afforded by surface treatments against the effects of carbonation. Early work was carried out in Germany by Klopfer[7.11] and Engelfried.[7.12] They carried out tests on surface treatments applied to paper or cardboard discs. The discs were used to seal small cup-like containers which contained a carbon dioxide absorbing chemical. The containers were then exposed to an atmosphere enriched with carbon dioxide. The amount of carbon dioxide passing through the paint film was monitored by measuring weight gain at various time intervals. As a result of this work it has been found possible to propose diffusion resistance requirements for surface treatments for concrete to provide protection against carbonation. Confusingly the diffusion resistance figures are often quoted as an equivalent thickness of air layer.

In the United Kingdom a joint programme by the Building Research Establishment and the Paint Research Association[7.13] has compared the permeability of coatings on mortar blocks and the same materials as free films or on paper backings. In the former case, the test specimens were 50 mm × 50 mm × 25 mm mortar blocks which were exposed for 14 days in a carbon dioxide-enriched atmosphere at relative humidities of

85 per cent and 60 per cent. During the test period the weight of the blocks was checked at intervals. At the end of the exposure period the blocks were split open and the carbonation depths measured using phenolphthalein.

In the other series of tests, the film was mounted in a test cell so that it separated the cell into two isolated chambers. A steady stream consisting of 30 per cent carbon dioxide at controlled relative humidity and temperature was passed through one chamber of the cell. A sweep stream of air free from carbon dioxide at similar temperature and relative humidity was passed through the other chamber. The sweep stream was sampled after it had passed through the cell and the concentration of carbon dioxide measured allowing the permeability of the coating to be assessed. When the results of the tests were compared it was found that the resistance to penetration determined by the two tests differed by factors ranging between 2.5 and 100 for the same coating. Resistance of the free films was generally found to be higher than those on the mortar blocks. The differences were attributed to dissimilarities in the roughness of the substrate and this conclusion serves to illustrate the potential dangers of attempting to apply the results of comparative laboratory research to practical situations.

A split cell was also employed in another series of tests in the United Kingdom reported by Robinson.[7.14] The test programme involved 71 different proprietary coating products which were applied to unglazed ceramic plates. In this case the test gas consisted of 15 per cent carbon dioxide and 85 per cent oxygen. The sweep stream was of helium at a balanced pressure and it was tested by gas chromatography to determine both carbon dioxide and oxygen diffusion coefficients. Permeability to water vapour was determined for the same coatings applied to a cartridge paper substrate using a test cup similar to that used by Klopfer as described above. The main findings were that the performance achieved by the coatings varied over a wide range even within a given generic type.

In general it was found that solvented and pigmented coatings gave good resistance to passage of carbon dioxide but that they also restricted the passage of water vapour. The ability to transmit water vapour is an important property for a concrete coating as it reduces the possibility of vapour pressure building up behind the coating and causing adhesive failure. Some water-based systems were found to have acceptable water vapour transmission rates and also relatively high resistance to the passage of carbon dioxide.

Cathodic Protection

Corrosion is an electrochemical process as discussed in Chapter 2. Anodic and cathodic areas develop on corroding materials or systems, with

(a)

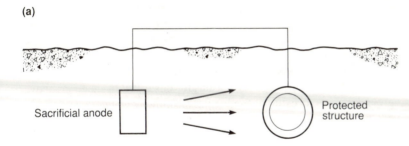

Sacrificial anode

Protected structure

(b)

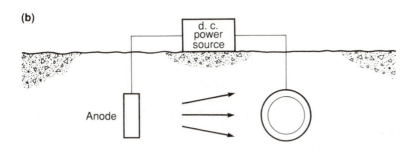

Anode

(c)

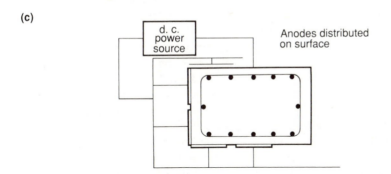

Fig. 7.5 Cathodic protection systems. (a) Sacrificial anode; (b) impressed current; (c) schematic for reinforced concrete

measurable potential differences between the anodes and the cathodes, and corrosion taking place at the anodes. It should therefore be possible to reduce corrosion rates if the whole of the material under consideration can be shifted to a cathodic condition by some externally applied potential. This, in very simple terms, is the theory behind cathodic protection.

Cathodic protection has been used successfully for many years by marine engineers to protect the hulls of vessels and also in the oil industry to protect pipelines and other important steel structures. There are two basic systems in use. In the first, often used in underground situations, the structure to be protected is connected to a more reactive metal as shown in Fig. 7.5(a). Anodes of aluminium, magnesium or zinc can be used in this way to provide protection to buried steel structures. The anodes corrode and are eaten away in the process of providing protection and for this reason they are known as sacrificial anodes. Although applications of sacrificial anode systems to prestressed concrete have been reported,[7.15] this method is not generally used for reinforced concrete structures. One of the reasons for this is that in situations above ground there is no external conducting medium by which return current can be distributed to the face of the structure to be protected.

The other system of cathodic protection works by the application of an external direct current electrical supply, as illustrated in Fig. 7.5(b). This is called the impressed current system. In the case of reinforcement in structures above ground, the current flows through the concrete and the anode cannot be placed further away than the surface of the concrete. As a result of this, and the high electrical resistivity of concrete which reduces the spread of current, it is necessary to use an anode system which is distributed across the surface, if all of the reinforcement is to be protected. The reinforcement also has to be electrically interconnected. A schematic cathodic protection system for a reinforced concrete structure is illustrated in Fig. 7.5(c).

Cathodic protection is promoted generally for the situation where reinforcement corrosion has come about because of the presence of high concentrations of chlorides in the concrete. It has not generally been put forward in the case of carbonated concrete because of the increase in electrical resistivity which occurs with carbonation and also because damage is often limited to a small proportion of the surface of the structure where cover is low. Conventional repair techniques can provide a durable and economical solution in such situations.

In the case of chloride-contaminated concrete this is rarely the case. It is not generally economic or possible to remove all contaminated concrete. If only cracked and spalled areas are repaired, a large area of reinforcement may remain in concrete with chloride concentrations sufficient to activate additional corrosion sites. Further cracking and spalling may occur. In these circumstances cathodic protection may provide the possibility of an economic means of prolonging the life of the structure.

Although the concept of cathodic protection is relatively straightforward there are many practical difficulties in its implementation, particularly in the case of application to reinforced concrete structures. Prior to its use on reinforced concrete superstructures, cathodic protection had principally been applied to structures immersed in an electrolyte such as sea water or soil. This allowed the anode to be located away from the surface of the member being protected and hence the protection current was able to spread and attain a degree of uniformity at the protected surface.

In the case of reinforced concrete superstructures there is no appropriate electrolytic medium. The reinforcement is surrounded by concrete which is capable of conducting electricity but the concrete is surrounded by air. This being the case, the anode or anodes have to be distributed over the concrete surface if a reasonably uniform cathodic protection current is to be supplied to the reinforcement.

The principal differences between the normal cathodic protection situation, where steel is in water or soil, and reinforced concrete have been outlined in a Concrete Society Technical Report[7.16] as follows:

1. The electrolyte, the pore fluid within the concrete, is contained within a rigid material which is highly alkaline and which is of low permeability to water and oxygen.
2. The anode has to be distributed over the surface of the structure and is closer to the steel being protected.
3. The nature of the cathodic polarization is different from steel in water or soil.
4. The protection criteria which have been established for steel in water or soil are not appropriate due to the different electrolyte.

These differences have meant that cathodic protection systems and particularly anodes have had to be developed specifically for the case of reinforced concrete. The original thrust for development came in the United States when engineers were faced with the problems of maintaining car park decks and unwaterproofed bridge decks exposed to de-icing salts. Work by Stratfull[7.17] resulted in the installation of a system on a bridge deck in 1958 and several others followed.

The early installations in the United States used cast iron anodes distributed over the deck in an overlay of asphalt containing coke to improve conductivity. The overlay was typically of 50 mm thickness but because of its poor physical properties it was necessary to provide an additional 50 mm wearing course. This was a significant addition to load and the system was disrupted when resurfacing was required. In order to overcome these difficulties, the next development was to install the anodes and cabling in slots cut into the deck.

The use of slotted anode systems was developed further by the use of platinum-clad niobium wire grouted into the slots using a conductive polymer mortar. The first installations were carried out in 1977. There were some difficulties with these initial installations caused by the high current densities made necessary by the small dimensions of the anode system within the slots. The high current densities led to production of acid which attacked the polymer mortar and adjacent concrete and caused some failures.

In the late 1970s and early 1980s there was much activity in the development of distributed anode systems which could be used on vertical surfaces and soffits. Developments have resulted in two main types of anode system. These are conductive coatings and wire or mesh anodes.

The first trials with conductive coatings used modified concrete as a secondary anode for a primary system of wires. Since then developments have tended towards paint-like conductive coatings. One of the most widely used has consisted of a solvent-based acrylic mastic containing graphite and it is understood that over $20\,000\,m^2$ have been applied in North America and elsewhere.[7.18] A large scale trial on a multi-storey office building using a conductive chlorinated rubber coating has been carried out in the United Kingdom.[7.19]

Distributed wire anodes have developed from the slotted anode systems using similar wires laid in mounds of the conductive polymer in a grid pattern. These were installed on the surface of bridge decks and multi-storey car-parks with a protective overlay of non-conductive concrete. Copper-cored conductive polymer wires had previously been used in soil and water, and their use was developed for reinforced concrete. They have been used extensively for bridge decks overlaid with conventionally placed concrete and also on vertical and soffit surfaces with sprayed concrete, since their introduction in 1983.

Anodes of titanium in expanded mesh form have been available in the United States and the United Kingdom since 1985. They are overlaid with concrete on horizontal top surfaces or with sprayed or hand-applied cementitious render on vertical surfaces and soffits.

Elements of System

A cathodic protection system consists of three main elements as follows:

(a) power supply and supply cables;
(b) anode including any overlay; and
(c) monitoring equipment.

Some factors governing the choice of these elements will be discussed for each element in turn.

Power Supply and Cables

In most cases cathodic protection systems are powered from a.c. mains supply and it is necessary to provide a transformer−rectifier to step down the voltage and to produce the d.c. supply that is required. It is often necessary to divide the structure into different zones to accommodate varying reinforcement densities or other variables and for this reason it is usual to use transformer−rectifiers which have multiple outputs which can be individually adjusted. Each output may be controlled to provide constant current or constant voltage with a limit on the voltage or current as appropriate. Each zone of the structure will require a power output typically in the range 40–150 watts with low current and low voltages typically in the range 5–50 volts. The output voltage has to be sufficient to provide for the reductions associated with the resistivity of the connecting cables, reinforcement and concrete, and the electrochemical potentials produced at the surfaces of the anode and the reinforcement whilst still providing the current density required for protection.

Cathodic protection installations may be carried out in extreme climates and it is necessary that transformer–rectifiers are adequately protected with weatherproof and vandal-proof cabinets. It may be necessary to make provision for cooling or to specify units which are capable of operating over the anticipated range of temperatures and humidities.

When designing the cabling and connection between the power supply and anodes and reinforcement, it is necessary to provide additional circuits to take account of the possibility of one or more breaks or failures. The connections to the reinforcement will require the concrete to be broken out and the bars cleaned. A positive connection must be provided and mechanical methods such as clamping or drilling and tapping have been used successfully, as have thermal methods such as brazing. The junction must be adequately protected by coating as the copper/steel combination gives potential for galvanic corrosion.

Interconnection between the reinforcement in the structure needs to be checked. Any reinforcement which is not connected directly to the system may become anodic under the influence of cathodic protection leading to an increased risk of corrosion. It is therefore important to carry out a check on the electrical continuity of reinforcement. If discontinuities are found they can be made good by breaking out and by providing leads interconnecting the reinforcement, using one of the methods mentioned above for junctions between reinforcement and supply.

Anode Systems

The history of anode development has been described in an earlier section of this chapter. In view of the fact that the anode or anodes are distributed across the surface of the member and that current has to pass through the concrete to the reinforcement, it is necessary to identify and deal with any areas of delamination before the anode system is installed. Delaminations are located by sounding the surface with a light hammer. Delaminated areas give off a hollow, dull thud in contrast to the ring given off by sound concrete. The surface should also be examined for the presence of tying wire or nails, which may provide a short circuit between the anode and the reinforcement. Suspect areas are broken out and reinstated using patch repair techniques.

The resistivity of the repair material is important as localized areas

Fig. 7.6
Conductive
coating cathodic
protection system

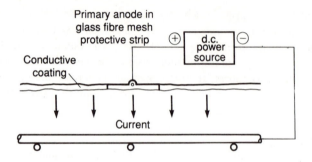

of high or low resistivity have an effect on the current density. In general, the resistivity of the repair material should be similar to that of the concrete in the member. Resin-bound materials, such as epoxy mortars, should not be used as they have a much higher resistivity than normal concrete. Most anode systems rely on the bond between the system and the concrete surface. Installation procedures for most anodes therefore include roughening of the concrete surface as well as the removal of any surface treatment which may affect the bond or the passage of current.

There are seven main generic types of surface anode in use in cathodic protection systems. A full description along with operating characteristics is given in an appendix to the Concrete Society report.[7.16]

Conductive Coatings

These are proprietary coatings containing a conductive filler such as powdered graphite. They are applied at a thickness of the order of 400 μm and utilize current feed-wires of titanium or carbon fibre mesh as shown in Fig. 7.6. The feed wires are spaced at roughly 3 metre centres. The coatings are dark in colour because of the graphite content, but they can be overcoated either for aesthetic reasons or to reduce temperature gain. As with most coatings designed for concrete, the life expectancy for conductive coating anode systems is anticipated to be of the order of 10–15 years. The electrolytic processes involved in cathodic protection may result in production of acid at the concrete/coating interface which would cause the bond to fail and reduce the life of the system. Local damage caused by impact or abrasion can be repaired easily and is not likely to cause significant problems as the current provided by adjacent areas of coating should continue to provide protection.

Titanium Mesh

In this case the anode is formed from titanium mesh produced by the expanded metal process. This results in a diamond-shaped mesh with strand thickness in the range 0.5–2 mm and apertures typically with aspect ratio of 1 : 2 with shorter dimensions in the range 30–100 mm.

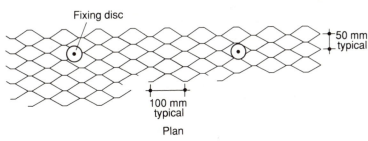

Fig. 7.7 Mesh
cathodic protection
system

The strands are given an application of a proprietary coating which differs
from manufacturer to manufacturer. The coating has an active electrocatalytic role in the cathodic protection circuit. Mesh anodes are designed
to be overlaid with a layer of cementitious mortar which encapsulates
the strand and passes current to the concrete.

Mesh can be cut easily to the required shape and can be bent to go
round corners or to cope with members of curved section. It is fixed
to the member with plastic fittings which are a push-fit into predrilled
holes in the concrete. The plastic fittings are designed to provide fixity
to the mesh and also to space the mesh at the desired distance from the
concrete surface. Connections to the supply are made by welding or
crimping and similar techniques can be used for strip connections between
adjacent areas of mesh. The cementitious overlay may be applied by
spraying or trowelling. Details of a completed system are shown in Fig.
7.7.

Mesh anodes have been designed to have a life expectancy in the range
10—15 years on the basis of the anticipated consumption rate of the active
coating. However, as in the case of conductive coatings, there is the
possibility of generating acidic products at the anode and these may cause
the life of the system to be curtailed. Protection also relies on an adequate
bond between the concrete surface and the overlay.

The mesh anode is not as easy to repair as conductive coating, but
it would be possible to repair very localized damage or areas of
debonding. If this is deemed necessary, it may be done by careful cutting
out and replacement using strip connectors to the new area of mesh.

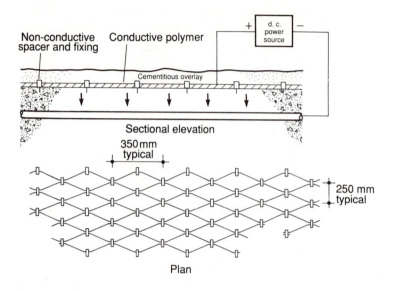

Fig. 7.8
Conductive
polymer cathodic
protection system

Conductive Polymer Anodes

These anodes are constructed from strand with a copper core surrounded by a proprietary conductive polymer which contains carbon. The strand is typically 8 mm in diameter and is laid in wave patterns on the concrete surface. It is held in position using purpose-designed clips of the same conductive polymer or of plastic. The system is designed to be encapsulated by an overlay material with a cover of the order of 25 mm. The thickness of the strand with clips is typically 12 mm. A completed system is shown diagrammatically in Fig. 7.8.

A manufacturer has claimed an anticipated operational life of 30 years at the recommended current densities. The electrochemical reaction at the anode may cause the carbon in the polymer to oxidize. This reaction progresses inwards and, if it reaches the copper core, fairly rapid attack on the conductor can result. As for any anode involving an overlay, the bond with the concrete substrate is of critical importance and this is another potential failure mechanism.

Sprayed Zinc

This system is similar in some respects to a conductive coating. A layer of metallic zinc is applied to the concrete surface at a thickness of some 200 μm. Application is by arc spray or oxygen/acetylene or oxygen/propane gun. Electrical connection is by titanium or stainless steel plates fixed to the concrete with an insulating layer of resin before spraying. The zinc coating may be overcoated with a conventional paint treatment for aesthetic and protective purposes.

Zinc is consumed in the cathodic process and the 200 μm thickness gives a life expectancy of the order of 10 years. Corrosion of the zinc may occur under wet conditions but the situation can be improved by overcoating. Any localized areas of damage due to corrosion, abrasion or some other physical cause can be repaired by local patching after suitable preparation of the surface.

Conductive Ceramic

These anodes are built up from tiles of a conductive ceramic material. Each tile is approximately 50 mm × 50 mm × 2.5 mm and the material from which they are made is resistant to acids and alkalis. The tiles are stuck onto the concrete surface using a modified cementitious grout, and the current is supplied through titanium conductor strips attached to the tiles using self-tapping screws and washers, also of titanium. The screws which connect the conductor strips to the tiles also penetrate into the concrete and provide an additional mechanical fixing. Plastic plugs are used in the concrete if there is a danger that the screws could contact the underlying reinforcement.

Conductive Overlays

These are used mainly on bridge or car-park decks and are formed using primary anodes of high-silicon cast-iron overlaid with a conductive asphalt. The cast-iron disc anodes may be installed in recesses in the concrete so that future resurfacing can be carried out without disrupting the system. The conductive asphalt is normally laid at around 40 mm thickness but may be covered by a conventional asphalt as a wearing surface.

The conductive asphalt is subject to the same rigours as any bridge overlay and will have to be replaced within 10−15 years depending on traffic loading and the environmental conditions of exposure. This system of cathodic protection for reinforced concrete has been in continuous use for a longer period than any other. Some of the earliest systems are reported to have given satisfactory service for over 10 years.

Conductive Resins

A Federal Highways Authority research programme in the United States has developed anodes of resins made conductive by the addition of graphite for use in cathodic protection systems on bridge decks. The resins have been used in two ways. In the first method, slots were cut to a depth of 25−30 mm in the deck and a primary anode of titanium or niobium wire was placed into the slot before filling with the conductive resin. In the second method, the resin anode system was placed in linear mounds on the deck before covering with a cementitious overlay. In both cases the anodes formed a grid pattern with the titanium or niobium wires in one direction at approximately 3 metre centres and the strips containing carbon fibres in the direction at right angles at approximately 300 mm centres.

Testing and Operation

The operation of the protection system has to be monitored during its lifetime and provision for this monitoring has to be at the time of installation. If portable copper/copper sulphate half-cells are to be used to monitor the potential of the reinforcing steel it will be necessary to leave windows in the anode and overlay so that contact can be made with the surface. An alternative is to use permanently embedded reference electrodes of the silver/silver chloride/potassium chloride type.

The locations for reference electrodes need to be carefully chosen so that they are representative, taking into account exposure condition and orientation of various faces. Additional sites may be chosen at locations of high potential identified in the pre-installation survey where corrosion is suspected. Locations in or close to repairs should be avoided since they are not representative of the main body of non-repaired concrete.

Pick-up probes are also used to measure the effectiveness of cathodic protection systems. They are also sometimes known as macro-cell probes and consist of a short length of reinforcement electrically connected to the main cage. The probe is embedded in the concrete using a repair mix and the current flowing along the connection to the main cage is measured. Before cathodic protection is applied, the probe is anodic and therefore there will be a current flow towards it. When the cathodic protection voltage is applied the current flow should be reversed. The probes therefore give a straightforward demonstration that the system is providing a cathodic protection current to that part of the structure.

Electrical resistance probes can also be used. These consist of small coupons of metal cast into the concrete with suitable connections so that their resistance can be measured. If corrosion takes place, the cross-sectional area of the coupon is decreased and as a consequence the electrical resistance increases. In cathodically protected structures the coupons are connected to the reinforcement and if the protection is working the change in resistance should be small.

The criterion by which effectiveness of cathodic protection is judged has been the subject of discussion and is not yet finally settled. Several possibilities have been suggested including:

(a) polarization of all reinforcement to a negative potential in excess of 770 mV with respect to a copper/copper sulphate reference electrode;

(b) potential decay of at least 100 mV;

(c) potential shift of 300 mV; and

(d) $E-\log I$ determination.

A joint working party of the Concrete Society and the Corrosion Engineering Association[7.16] has recommended the use of the 100 mV potential decay criterion. To measure the decay potential the power supply to the anode is switched off. The 'instantaneous off' potential is measured at a time between 0.1 and 1.0 seconds after interruption of the power supply. This 'instantaneous off' potential is subject to a most negative limit of -1.15 volts with respect to copper/copper sulphate. The cathodic protection system remains switched off and the measured potentials gradually decay. A minimum of 100 mV potential decay is required over all representative points. The rate at which decay takes place is dependent on the rate at which oxygen is able to diffuse to the reinforcement. Hence the time for depolarization will depend on the permeability of the concrete and its degree of saturation, and may vary across a structure and between structures. Periods of 4 hours have been reported for bridge decks in the United States, but longer periods in excess of 1 day may be appropriate in some cases, particularly for structures which have been protected for a considerable period or which are saturated.

Chloride Extraction and Replenishment of Alkalis

Patented techniques, akin to cathodic protection, have been developed as once-off treatments for chloride-affected and carbonated reinforced concrete.[7.20] The processes involve the application of an external potential which causes chloride ions to be moved away from the reinforcement and hydroxyl ions, which increase alkalinity, to be generated at the reinforcement. As in the case of cathodic protection, a distributed anode is mounted on the surface with an overlay. In this case, the overlay is of sprayed cellulose fibre saturated with an alkaline solution. Chloride ions migrate towards the anode and are removed with the overlay.

As with the case of cathodic protection, there is no need for wholesale breaking out and reinstatement before application of the technique. However, any loose or delaminated areas have to be repaired as the process relies on a continuous medium between the reinforcement and the surface anode. It is also necessary to treat any features such as tying wire, exposed reinforcement or cracks which extend to the reinforcement as they could cause short circuits between the steel and the anode, and significantly reduce efficiency. Paint or other coatings must be removed as they would have an insulating effect.

To install a system, electrical connections are first established with

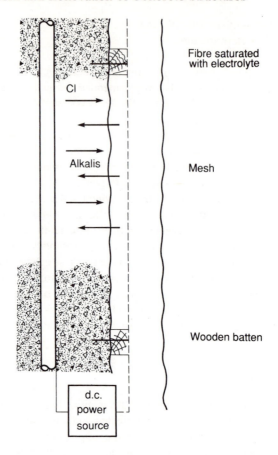

Fig. 7.9 Chloride extraction and replenishment of alkalis

Fibre saturated with electrolyte

Cl

Alkalis

Mesh

Wooden batten

d.c. power source

the reinforcement. Typically one contact point is made for each 50 m^2 of concrete surface. Wooden battens are attached to the concrete surface using plastic fixings and a layer of cellulose fibre saturated with electrolyte is applied by spray. The temporary anode is installed on the wooden battens using plastic plugs and bedded into the fibre layer. Anodes are usually in the form of mesh which can be easily bent and cut to shape. Mesh may be of titanium or steel. Titanium mesh is inert and can be reused; steel mesh corrodes during the application of current and may cause stains on the surface which are difficult to remove. Finally, a further layer of cellulose fibre is sprayed over the mesh and connections are made to a d.c. power source. Connection is usually made to the anode every 10 m^2. A completed installation is shown diagrammatically in Fig. 7.9.

The voltage is applied such that the external mesh becomes the positively charged anode and hence attracts the negatively charged chloride ions. They move through the concrete and into the fibre layer.

At the same time, hydroxyl ions are produced at the negatively charged reinforcement thus increasing the pH at this location.

$$2H_2O + O_2 + 4e^- = 4OH^- \qquad [7.1]$$

The alkaline solution in the fibre layer may be drawn into the concrete by a process of electro-osmosis, again increasing the pH. The electrolyte used in the fibre is usually a 1 molar solution of sodium carbonate. If alkali aggregate reaction may be a problem, calcium hydroxide can be substituted.

Replenishment of the alkalis is a quicker process than extraction of chlorides. The former process may only require a few days while the latter can take several weeks. Extraction of chlorides is more difficult because some of the chloride may be bound up in reaction products with the cement hydrates. The bound chlorides are in dynamic equilibrium with the chlorides dissolved in the pore water. As the chlorides in solution are removed by the electrolytic process, they are replaced by chlorides from the reaction products. The rate at which chlorides can be extracted is controlled by the rate at which the decomposition of the hydrates occurs. To overcome this effect, current has been applied intermittently in some cases. The rest period allows time for more chloride to be liberated from the reaction products and these are extracted when the current is reapplied giving an increase in the overall efficiency of the operation.

The processes of chloride extraction and replenishment of alkalis are relatively new, but it has been reported that the technique has been applied on structures in Europe, the Middle East and Hong Kong. Also tests taken after application show substantially reduced chloride concentrations and a significant increase in pH.

References

7.1   Vassie P 1984 Reinforcement corrosion and the durability of concrete bridges. *Proceedings of the Institution of Civil Engineers, Part 1* **76** **August:** 713–23.

7.2   CIRIA 1987 *Protection of Reinforced Concrete by Surface Treatments* Technical Note 130, The Construction Industry Research and Information Association, London

7.3   British Standards Institution 1966 *Glossary of Terms Applicable to Internal Plastering, External Rendering, and Floor Screeding* BS 4049, The British Standards Institution, London

7.4   Roth M 1982 Silicone masonry water repellents for the surface impregnation of mineral building materials. *Baugewerbe* **82 (2)**

7.5   Rodder K M 1977 Protecting buildings by impregnation with water-repellent silanes. *Chemische Rundschan* **22**

7.6   Hammersley G P, Dill M J, Darby J J 1990 An investigation into the effectiveness of silane for reducing corrosion activity in a chloride-contaminated reinforced concrete bridge structure. *Proceedings of the International Conference on Bridge Management* University of Surrey, Elsevier, London, pp 655–66

7.7   Gaul R W 1984 Preparing concrete surfaces for coatings. *Concrete International* **July:** 17–22

7.8   Department of Transport Highways and Traffic 1990 *Criteria and Material for the Impregnation of Concrete Highway Structures* Departmental Standard BD43/90 Department of Transport, South Ruislip

7.9   British Standards Institution 1975 *Methods of Test for Paints. Determination of Film Thickness* BS 3900: Part C5, The British Standards Institution, London

7.10   National Cooperative Highway Research Program 1981 *Concrete Sealers for Protection of Bridge Structures* Report 244, The Transportation Research Board National Research Council, Washington DC

7.11   Klopfer H 1978 Carbonation of exposed concrete and its remedies. *Bautenschutz Bausanierung* **1 (3):** 86–97

7.12   Engelfried R 1983 Diffusion resistance coefficients for carbon dioxide and water and their applications.*Farbe und Lack* **89 Jahrgang 7:** 513–18

7.13   Treadaway K W J, Rothwell G W 1987 Coatings for concrete buildings. *Paint Research Association Symposium on Protection of Concrete and Reinforcement*, May, pp 18–23

7.14   Robinson H L 1986 Evaluation of coatings as carbonation barriers. *Proceedings of the Second International Colloquium on Materials Science and Restoration* Technische Akademie, Esslingen

7.15   Cherry B W 1989 Cathodic protection of underground reinforced concrete structures. *Institution of Corrosion Scientists and Technologists Second International Conference on Cathodic Protection Theory and Practice*, June

7.16   Concrete Society 1989 *Cathodic Protection of Reinforced Concrete* Technical Report No 36, The Concrete Society, Wexham

7.17   Stratfull R F 1959 Progress report on inhibiting the corrosion of steel in concrete bridge decks. *Corrosion* **15:** 65–8

7.18   Wyatt B S, Irvine D J 1987 A review of cathodic protection of reinforced concrete. *Materials and Performance* **26 (12):** 22–8

7.19   Pook C, McAnoy R 1988 Current experience in the application of cathodic protection to buildings. *Seminar Cathodic Protection of Concrete Structures — the Way Ahead* The Institution of Structural Engineers, London, pp 21–30

7.20   Anderson G 1990 Chloride extraction and realkalization of concrete. *Hong Kong Contractor*, **July–August:** 19–25

8 Contract Documents

Introduction

There are a number of inherent difficulties in drawing up contract documents for concrete rehabilitation projects. The difficulties stem mainly from the uncertainty surrounding the extent and, in some cases, even the nature of the work to be undertaken. In most cases an investigation and survey will have been carried out using methods similar to those described in earlier chapters and the information will be available at the time when documents for remedial work are being drawn up. The investigation will have determined the nature of the deterioration and its causes and the present condition of the structure. This information will permit the general form of the repairs to be designed. However, it is unlikely that the survey will have been sufficiently detailed to locate all of the defects. This is because only limited access will have been available for the survey and also because some defects may have been hidden by coatings or by surface dirt.

A further difficulty in drawing up documents is that the surface indications of damage such as cracking or spalling do not give an adequate guide to the extent of individual repairs either in terms of area or depth. As an example, cracks due to corrosion of reinforcement may not extend over the full length of the rusted portion of the bar and repairs will be much more extensive than the surface crack. There may be a considerable variation in the cover to reinforcement over the area of a structure and this will have a bearing on the depth of individual repairs.

For these reasons, documents for concrete rehabilitation contracts have to be drawn up in a way which permits a flexible approach. This has the disadvantage that the cost of the work is not known at the start of the contract and the owner of the structure should be made aware of this situation so that budget reserves can be allowed. A second consequence is that few rehabilitation contracts are drawn up on a fixed price basis and, in most cases, remeasurement is permitted.

The elements making up the documentation for the rehabilitation of a reinforced concrete structure are generally similar to those found in most construction contracts in that they consist of:

1. Conditions of contract.
2. Drawings and reports.
3. Specification for the work.
4. Bill of quantities.

Conditions of Contract

Standard conditions of contract have been drawn up by several of the bodies connected with the construction industry in the United Kingdom. These include the Institution of Civil Engineers[8.1] and the Joint Contracts Tribunal.[8.2] A form of contract for large projects abroad has been published by FIDIC.[8.3] These forms were drawn up for new work but many have also been used on rehabilitation contracts with only minor amendments. However, lump sum forms of contract are only applicable to small repair contracts and in general some provision for remeasurement is necessary for the reasons outlined above.

Drawings and Reports

The prospective contractor should be provided with all information which will be relevant and helpful in pricing the work. The drawings should include the normal site locations and layouts. Reinforcement drawings from the time of construction are also useful if they are available, as they indicate the reinforcement content and the cover that was intended. However, it should be made clear that there may have been changes in the time between the production of the drawings and construction, and also that the cover achieved at site may have been extremely variable.

Information relating to any phasing requirements can also be conveniently shown on drawings. In the case of an occupied building, or other structure which remains in use during rehabilitation, locations at which access must be maintained can also be shown. Special measures may have to be taken at these positions to provide protection for the users of the building.

For the reasons stated earlier, it is unlikely that it will be possible to prepare drawings which indicate the location of each defect requiring repair. It may be possible to produce drawings which indicate the nature of typical faults and the probable repair types.

If possible, copies of any reports which have been prepared describing the condition of the structure and the causes of defects should also be made available to the prospective contractor.

Specification for the Work

Standard specifications have been produced which cover several different aspects of concrete rehabilitation work. These include patch repair,[8.4] sprayed concrete[8.5] and cathodic protection.[8.6] The subjects which need to be addressed in a specification include access, surface cleaning, surveys during the contract, breaking out, reinstatement, materials and their use and testing. Several of these subjects are independent of the type of work being carried out.

When drawing up the specification, one of the first matters that needs to be decided is who is to undertake the survey of the defects during the contract and make the choice of the type of repair to be implemented. Two different approaches are in common use. In the first, the contractor undertakes the major part of the survey work and marks up each of the repairs which are then agreed by the Engineer. The specification or drawings must give the basis for the decisions if a variety of defects and repair types are anticipated. The alternative, which requires that the Engineer has a much greater involvement at site, is that the Engineer undertakes the survey work and marks up the areas for repair.

The specification should make it clear as to which of these approaches is to be adopted as the choice has significant programming implications. It can be useful if the specification includes a description of the anticipated sequence of events including times when the Engineer requires access for survey, approval or measurement.

In other areas, the specification should not be unnecessarily restrictive. The choice of means of access or method of surface cleaning can, in many cases, be left to the contractor if the general requirements are given in the documents.

Although it is difficult to cover all cases likely to be encountered in concrete rehabilitation works, the following headings may be found useful when drawing up a specification:

General
Access
Cleaning
Survey
Breaking out
Reinstatement
Surface protection
Materials
Trials and testing

The General section will differ from project to project. It should give information particular to the structure under repair and specific to the contract. There should be clauses describing the structure and the nature of the defects. Accompanying documents, such as drawings and reports, should be listed. This section can also be used to draw the contractor's attention to any restrictions on working hours or the timing of noisy operations. The intended sequence of operations should be set out.

As discussed above, the choice of the actual form of access (scaffolding, cradles, self-elevating platforms, etc.) in such circumstances may be best left to the contractor as this leads to greater flexibility and hopefully economy. The contractor should be asked to state the chosen form of access at the time of tender. The specification should address matters such as fixings and loadings on the structure from the access equipment. Any requirements for enclosing the work area should also be stated. It may be necessary to state any additional access requirements either ahead of the work to permit surveys and trial repairs to be undertaken, or at the end of the defects liability or maintenance period to allow the Engineer to carry out inspections.

Cleaning of the surface can be specified either by method (water jetting, grit blasting, etc.) or by stating the required end result. The latter alternative again permits the contractor a free choice of method and may consequentially lead to overall economy.

Some form of investigatory and survey work will have been carried out prior to the drafting of the contract documents to allow the appropriate form of remedial works to be designed. However, there will be a need for a comprehensive survey during the contract to identify all of the defects requiring repair. As stated earlier in this chapter, the survey may be undertaken by the Engineer administering the contract or by the contractor.

In the former case, the specification must state the time requirements for the Engineer to complete the survey and mark up the defects, after access is in-place and the surface has been cleaned, so that the contractor can allow for this in the programme. The specification should also indicate any services or equipment which has to be supplied by the contractor through the contract.

In the latter case, the specification should cover the qualifications of the personnel undertaking the survey, the frequency with which measurements, such as covermeter readings, are to be taken and what type of records are to be kept. Clauses will also have to be provided to give the contractor sufficient information so that there is a rational basis for choosing the appropriate repair method based on the findings of the survey. As an example, if the precontract investigation had found that the structure was suffering corrosion of reinforcement as a result of carbonation and low cover, the specification might require that all cracked, spalled, hollow sounding and low cover areas are repaired.

The choice of method of breaking out may generally be left to the contractor unless there are constraints on noise or vibration. The main points that need to be brought out in the specification are that there should be no feather-edges (i.e. the repair should have a clearly defined edge and not just taper out), the minimum of concrete should be broken out and that the method of breaking out should not damage concrete that is to remain in-place. It may also be necessary to give limits on the amount of concrete that can be broken out at any one time in order not to impair the stability of the structure. Requirements for any temporary propping should be described.

There are three principal areas which have to be covered in the reinstatement section of the specification. These are surface preparation, cleaning of reinforcement and the application of corrosion inhibitor, and making good with mortar or concrete. The contents of each of these areas will depend on the types of materials being used and, in many cases, the instructions of materials manufacturers. Surface preparation will cover such matters as cleanliness, the need to presoak the surface if cementitious material is being used and the application of bonding aids or primers. It is usual to specify grit blasting as the method of cleaning reinforcement but also to state the quality that is to be achieved. There is a need to state a time limit between cleaning and application of corrosion inhibitor, particularly if the work is being carried out in a corrosive environment. Another point that needs to be addressed is the requirement for replacing or supplementing reinforcement if there has been severe loss of cross-sectional area because of corrosion. Clauses on making good should include any restrictions on the ambient temperature or the temperature of the substrate during reinstatement. They should also include limitations on individual layer thickness for hand applied repair mortars. The requirements for compaction and surface finish should be stated along with a statement on permissible curing methods.

The surface protection section of a specification should address surface preparation as well as the application of coatings. Surface preparation might include grit blasting to open up any blowholes, and the application of a filler. Specification clauses on coatings should include restrictions on application during unsuitable weather conditions and controls on application rates so that the required dry film thicknesses are achieved.

Materials clauses are difficult to draft at the present time because there are few national standards which cover this topic. One approach is to specify a number of repair systems from different manufacturers based on a successful track record on similar projects. Other clauses should cover aspects such as storage, mixing and any restrictions on composition, e.g. chloride and alkali content.

The lack of national standards also causes difficulties in producing the testing section. It is wise to include some routine strength testing for quality control purposes and fortuitously there is a suitable available standard. British Standard 6319[8.7] includes a strength test on a 40 mm cube.

The pull-off procedure could also prove to be useful for testing repairs and a Dutch CUR Recommendation[8.8] is applicable. In this method, as described in Chapter 4, a 50 mm diameter core is cut through the repair and an axial force is applied. The force required to fracture the core is measured. The test is useful because as well as giving the pull-off strength, it also permits an assessment of the degree of compaction achieved in the repair material by visual inspection of the core. The position of the break is also important. The break should not occur on the bond line if the surface has been correctly prepared.

The specification should clearly identify the repairs which are represented by a particular set of test results, to try to avoid disputes about which repairs are at risk should a test fail. This is usually effected on a time basis. For example, if one set of tests is carried out each day the results would be considered to be representative of all repairs carried out on that day.

It is useful to specify that trial repairs are carried out early in the contract as these are a good test of both the materials and the operatives. Some of the trial repairs can be broken out to check the degree of adhesion and compaction that has been achieved. Others can remain in-place to provide a standard against which the remainder of the work can be judged.

Concrete Society Patch Repair Specification

As stated earlier, several standard or model specifications are available for different aspects of concrete rehabilitation. The Concrete Society document on patch repair[8.4] covers most of the topics mentioned above. A method of measurement is also given and a specimen bill of quantities. The introduction to the method of measurement highlights the difficulties of drawing up precise quantities at the time of tender and suggests that the bill is more akin to a schedule of rates. The method of measurement attempts to separately identify time-related items such as site set up and access to allow these to be easily reduced or increased if the amount of repair work proves to be significantly different from that shown in the bill.

The method of measurement has four sections:

1. General items which include site set-up, access, trial sample repairs and routine testing.
2. Surface cleaning and inspection and testing.
3. Repair.
4. Protection.

The sections for the work items are all subdivided according to the orientation of the surface — vertical, top surfaces and soffits.

Time-related elements in the General Items section of the bill are entered under three items — provide, maintain and remove. The maintenance item is measured on a time basis while the other two are measured as items. This method permits payment for these elements to be readily increased or decreased on a rational basis if there are changes in the amount of work. Routine tests, such as cube tests, are measured by number and trial sample repairs as a single item.

Surface cleaning, many types of survey, application of fairing coat and protective treatment all lend themselves to measurement by area. Testing, such as carbonation and dust sampling, is measured by number. Additional item descriptions are given for preparation of drawings showing results of survey and testing.

Repairs are generally measured by number in bands of depth and area, e.g. all repairs in the depth range 20−50 mm and the area range 0.1−0.25 m^2 are grouped together for measurement purposes. Repairs containing an external corner are separated from repairs to flat surfaces to take into account the additional cost of forming the arris.

Dutch Repair Recommendations

The Centre for Civil Engineering Research and Codes in Holland has issued recommendations for concrete repairs with polymer-modified cement mortars.[8.9] The recommendations are perhaps the closest approach to a national standard or code of practice currently available. The document gives definitions, classifications, requirements and standards for the composition and manufacture of polymer-modified cement mortars, and the execution, inspection and testing of repairs on concrete carried out manually or by the wet spraying method. It is assumed that an assessment of the structure has been carried out previously and that this has resulted in a requirement for repair using polymer-modified cement mortar.

The scope of application of the recommendations is stated as:

1. Restoration of the alkaline environment around the reinforcement to reinstate envisaged durability in cases of reinforcement corrosion.
2. Provision of cover to reinforcement with adequate density and thickness.
3. Levelling and repairing damaged areas of concrete surfaces.

The recommendations are in 13 sections dealing with a comprehensive range of topics including scope and definitions, material requirements, preparation, supply of materials, testing, and health and safety. Five of the sections are of particular interest and the contents are discussed below.

Classification

Mortars are classified according to the environment in which they are to serve and details of the particular application, e.g. does the repair area contain reinforcement. The environmental exposure conditions are given in a separate standard and are numbered in increasing severity from 1 to 5 with subdivisions. There are three main application grades designated Rc1 to Rc3, where the R stands for repair mortar and the c signifies cement-based.

Grade Rc1 mortar is suitable for repair and levelling of concrete surfaces exposed to the lower two environmental conditions and repairs involving exposed reinforcement in the least severe environmental condition. The mortar is to provide a suitable substrate for the subsequent application of coatings and is required to possess the properties listed in Table 8.1.

Table 8.1 Requirements for application Grade Rc1 mortars according to CUR Recommendations 21 (after Gouda[8.5])

Maximum air content	10%
Minimum characteristic compressive strength	15 N mm^{-2}
Minimum characteristic flexural strength	4 N mm^{-2}
Minimum characteristic bond strength	0.6N mm^{-2}
Minimum degree of compaction	95%
Maximum halogen content	0.05%
Maximum shrinkage	12×10^{-4}
Maximum coefficient of thermal expansion	15×10^{-6} °C^{-1}

Mortar in Grade Rc2 is suitable for repair and protection of reinforcement in exposure to Grade 2 environmental conditions or above. The basic requirements are as for Grade Rc1 but with additional requirements for water penetration, resistance to frost and de-icing salts, and resistance to carbonation. The actual requirements for these properties depend on the environmental exposure conditions indicated in Table 8.2.

Table 8.2 Additional requirements for application Grade Rc2 mortars according to CUR Recommendations 21[8.5]

	Environment grade			
	2a and 5a	5b	3,4,5c and 5d	
Maximum water penetration (mm)	30	20	10	
Frost/de-icing salt resistance				
Number of frost/de-icing cycles	5	10	15	25
Maximum loss of mass (mg mm^{-2}) (cumulative)	0.1	0.2	0.3	0.4
Maximum carbonation depth (mm)			2	

Grade Rc3 mortar has similar requirements to Grade Rc2 but with increased strength properties. Higher values can be set for characteristic compressive strength, characteristic flexural strength and characteristic bond strength, and the values have to be agreed prior to commencement of work.

Materials

Reinforcement, cement, aggregates, water, fillers and admixtures are all specified according to Dutch national standards for reinforced concrete. Four basic requirements for polymers are stated as follows:

1. Resistant to alkaline environment.
2. Shall form a closed pore structure.
3. Must not cause foaming or an antifoaming agent must be added.
4. Minimum film-forming temperature not greater than 5 °C.

The minimum permitted moisture content for preblended materials is given as 0.5 per cent by mass.

Composition and Properties

In addition to the properties listed in Table 8.1, limitations are placed on the maximum aggregate size, cement content and polymer content. Aggregate size is specified in accordance with certain dimensions within the repair. The maximum aggregate size in the mortar must not exceed the smaller of:

(a) one-third of the application layer thickness;
(b) one-third of the cover to be supplied to the reinforcement;
(c) one-quarter of the bar spacing; or
(d) one-quarter of the distance between the back of the bars and the substrate.

Minimum cement content is specified according to Dutch standards for concrete in relation to various environmental grades and the polymer content is required to be in the range 5−20 per cent by mass of the cement.

Preparation and Application

There are eleven subsections dealing with matters from qualification of personnel employed on the works to curing and measures to be taken during occurrence of low outdoor temperatures.

Requirements for personnel are that they are to be 'sufficiently skilled expert craftsmen'. A note explains that this means that the workers in question have passed a proficiency test or that they can demonstrate that they have carried out concrete repairs satisfactorily in the past.

The subsection dealing with preparation of the repair gives guidelines for the extent of concrete which is to be broken out. At locations of corrosion the concrete is to be removed to such a depth that non-carbonated concrete is reached. Breaking out continues along the reinforcement until non-carbonated concrete is encountered over a length equal to the required cover thickness. The width of the repair is to be such that a clear gap of at least the diameter of the bar is to be achieved on each side of the bar. Concrete is to be broken out in such a way that the surface of the substrate is left rough, there are no abrupt changes in the thickness of the repair and there is an angle of at least 60° between the side of the repair and the original concrete surface.

Restrictions are placed on reinstatement of repairs during adverse weather conditions. These are written in terms of the temperature of the substrate and the ambient temperature in relation to the minimum film-forming temperature of the polymers contained in the mortars. Additional restrictions relate to work in rainy weather when there is a risk that the quality of the repair or the finish may be compromised.

Testing and Checking

Details are given for so-called primary and execution testing. Primary tests are for use on repair mortars at the production stage to show that they meet the required standard. This testing may be carried out by an independent certification institute, in which case it is required once only unless the composition is changed. Alternatively, the testing is instituted and agreed by the client or purchaser and the supplier or repair firm.

Execution tests are used on material produced at-site to check that work has been carried out properly. This again may be carried out by an independent certification authority. The frequency of carrying out execution tests is related to the number of workers actually involved with the application of mortar.

Air Content

Air content determinations are carried out on three samples of wet mortar taken from three separate mixes for the primary test and at least once every five consecutive repair days for the execution check. Checks on air content during execution are not required for prebatched materials.

Preparation of Hardened Mortar Specimens

The remainder of the recommended tests are carried out on prisms, cubes, remnants of prisms after other tests or on composite test specimens. For repair mortars with maximum aggregate size below 8 mm the prisms are of dimensions 40 mm × 40 mm × 160 mm and the cubes are of 100 mm size. Composite test specimens are prepared by applying a 25 mm layer of mortar to a 300 mm × 300 mm × 50 mm concrete slab using a priming agent if this is to be used in the works. Specimens are subjected to various conditioning regimes according to the purpose of the test. The conditioning regimes are as follows:

Condition A. In a room at a controlled temperature of 20 ± 2 °C and controlled relative humidity of at least 95 per cent for 7 days followed by storage in conditions of 20 ± 2 °C and relative humidity $65 + 5$ per cent thereafter.

Condition B. As Condition A for the first 28 days followed by conditions which maintain the same temperature and relative humidity but in which the concentration of carbon dioxide is also controlled to 0.03 ± 0.003 per cent.

Condition C. As Condition A for the first 28 days followed by 28 days in saturated calcium hydroxide solution.

Flexural Strength
The specimens are prisms stored under Conditions A for 28 days. For the primary test three specimens are tested and for the execution check three specimens are made for each worker for each 5 days on which repairs are carried out. The results are assessed in groups of 12 and the acceptance criterion is that

$$x_m - 1.53S \geq f_f \qquad [8.1]$$

where x_m is the mean flexural strength of the 12 prisms;
$\quad S$ is the sample standard deviation of the results; and
$\quad f_f$ is the required characteristic flexural strength.

In the case where less than 12 results are available, the mean flexural strength has to exceed the required characteristic flexural strength by at least 3 N mm^{-2} and in addition no prism should give a result less than 0.9 times the required characteristic flexural strength.

Compressive Strength
Compressive strength is measured on one of the remnant parts of each of the prisms tested for flexural strength. The number of tests required is the same as for flexural strength. Where series of 12 results are available, the acceptability criterion is also the same as for flexural strength. Where less than 12 results are available, the requirement is that the average strength exceeds the required characteristic compressive

strength by at least 8 N mm^{-2} and that no result shall be less than 0.9 times the required characteristic strength.

Halogen Content

The halogen content is determined on remnants of prisms after carrying out flexural and compressive strength tests. The test is carried out on three separate samples in the primary test and no value should exceed the limit of 0.05 per cent by weight. No tests are required during execution.

Carbonation Depth

Testing is carried out on three prisms after storage under Condition B for the primary test. No carbonation depth testing is required during execution. The measured value for each prism is taken as the average of the value of carbonation depth on each side of the prism. In order to pass the test the average carbonation depth for the three prisms must not exceed 2 mm.

Shrinkage

This test is carried out by measuring length change on each of three prisms stored under Condition A. After 90 days the shrinkage must not exceed 12×10^{-4} on any of the specimens. Only primary tests are necessary.

Thermal Expansion Coefficient

Coefficient of thermal expansion is measured on each of three prisms after storage under Condition A for 28 days. The specimens are heated to a temperature of $80\,°C$ and the change in length is measured accordingly. An average coefficient for the three specimens of not greater than $15 \times 10^{-6}\,°C^{-1}$ is required. Only primary tests are required.

Water Penetration

For mortars with aggregate sizes less than 8 mm, the water penetration test is carried out on cubes after storage under Condition A for 28 days. In the case of a primary test three cubes are used; in the case of tests during execution of the work, one test is carried out for each five consecutive repair days but at least one test must be carried out per week. The cubes are required to have water penetrations not less than the values shown in Table 8.2.

Alkali Resistance

The test for mortar is carried out by comparing the flexural and compressive strength of three prisms stored under Condition C with values obtained from three prisms stored under Condition A for 56 days. The average of the three values of flexural strength for specimens stored under Condition C must be at least 75 per cent of the average value for those stored under Condition A. Similarly, the average value of compressive strength should be at least 90 per cent of the average value under Condition A. No determination is required at the execution stage.

For priming agents, the assessment is carried out by comparing the average bond results of three tests carried out on a composite specimen stored under Condition C with the average result from a specimen stored for 56 days under Condition A. The average bond strength for Condition C must be at least 90 per cent of the average value for Condition A. Only primary testing is required.

Bond Strength

For the primary test three determinations are made on each of three composite specimens after storage under Condition A for 28 days. The results are assessed according to the following criterion:

$$x_m - 1.67S \geq f_b \qquad\qquad [8.2]$$

where x_m is the mean value of the nine results;
$\quad\quad$ S is the sample standard deviation of the results; and
$\quad\quad$ f_b is the specified characteristic bond strength.

For tests during the execution of the work, the determinations are made on repairs in-place or, if this is not possible, on composite specimens as described above using mortar produced for the works. One test is carried out for each three repair days and the results are assessed in groups of 12. The acceptance criterion is:

$$x_m - 1.53S \geq f_b \qquad\qquad [8.3]$$

where the symbols have the same meaning as in Eq. 8.2. When less than 12 results are available, the mean strength should exceed the required characteristic bond strength by at least $0.7 \, \text{N mm}^{-2}$.

Resistance to Frost/De-icing Salts

The test is carried out on three composite specimens after storage for 28 days in Condition A. The loss in mass is measured after the mortar side of the specimen is subjected to various numbers of cycles of frost and de-icing. Measured loss in mass must be less than the figures indicated in Table 8.2. Only primary testing is required.

Degree of Compaction

The compaction achieved in the works is compared with that of the compaction achieved in the site-produced prisms used for the execution check on flexural strength. This is done by measuring the dry density of the mortar portion of cores from bond strength tests and of the remnants of prisms. The density determination is undertaken on one-third of the cores which have a remaining mortar layer of at least 20 mm. The requirement is that the cores shall, on average, have at least 95 per cent of the density of the prisms.

Sprayed Concrete

The Concrete Society have published a specification[8.5] and guidance notes on the measurement[8.10] of sprayed concrete or gunite. The specification is generally based on the performance principle. It sets the desired standard for the end product rather than specifying the details of the processes and materials which the contractor must employ. None the less, there is a materials section in the specification which gives the requirements for cement, aggregate and reinforcement to appropriate British standards. The detailed design of the mix is left to the contractor but it has to meet strength requirements stated by the Engineer.

The specification requires test panels to be sprayed prior to the commencement of the main works using similar plant to that proposed for the works. Cores are cut from the test panels and tested for compressive strength and examined for defects such as lack of compaction, dry patches, voids or sand pockets.

Details of preparation treatments to existing surfaces such as rock, concrete and brickwork are given along with the penetration resistance required before a subsequent layer of concrete can be sprayed onto a previously placed layer.

Spraying procedure is covered in three sections. These cover fixing and cover to reinforcement, the use of guides, rebound, construction joints, surface finishes, tolerances and curing.

The testing sections of the specification require further test panels to be sprayed during the course of the works. There is also an option for cutting cores from the test panels for crushing for determination of strength. Cores may also be taken from the works for strength testing or to check thickness and quality.

The Concrete Society *Guidance Notes on the Measurement of Sprayed Concrete*[8.10] is a straightforward document giving a schedule of 32 typical items from which the items for the bill of quantities of a particular contract can be chosen. Sprayed concrete is measured either by the cubic metre or by the square metre where the average thickness is stated. A preamble to the schedule gives details of the operations and materials included and covered by each of the items.

Department of Transport Specifications

The United Kingdom Department of Transport has published two standards dealing with repair[8.11] and impregnation[8.12] of bridges and

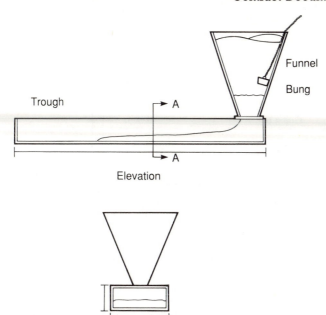

Funnel

Bung

Trough

A

A

Elevation

Section A–A

Fig. 8.1 Flow trough and funnel, after Department of Transport standard[8.11]

other highway structures. Departmental Standard BD27/86 gives model specification clauses for flowing concrete for use in repair, sprayed concrete and repair mortar. Two types of flowing concrete are described for different repair applications. One is a site-batched mix for use on decks and vertical surfaces of piers, columns and abutments. The other is for a proprietary prebatched concrete for use on the sides and soffits of beams and cross-heads. Sections of the specification cover storage and delivery of materials and site-mixing, placing and curing.

Details of two flow tests are described. The first uses a galvanized steel trough and funnel, as shown in Fig. 8.1, in which 6 litres of concrete are placed in the funnel and the bung is removed. The length of the flow along the trough is measured. The test is considered to be successful if the mix flows at least 450 mm along the trough without signs of segregation or bleeding. The second test involves a roughened precast concrete base slab and a reinforced section of 50 mm depth with a glass top plate as shown in Fig. 8.2. The concrete slab is wetted for 2 hours before the test. The mix is introduced at one end of the mould and is poured until it reaches a level at least 10 mm above the underside of the top plate at the other. The top plate is removed 24 hours later and the top surface of the pour is lightly brushed with a stiff bristled brush. The mix passes the trial if the formed concrete is homogeneous and free from air pockets.

The sprayed concrete and repair mortar sections of the document give

Fig. 8.2 Flow test
for horizontal
surfaces from
Department of
Transport
standard[8.11]

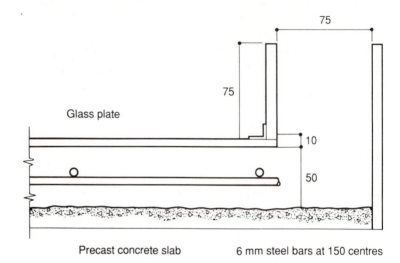

Fig. 8.2 Flow test for horizontal surfaces from Department of Transport standard[8.11]

Glass plate

Precast concrete slab   6 mm steel bars at 150 centres

short specifications for materials used for these applications. The subject
areas covered are similar to those in the materials sections of the Concrete
Society documents discussed earlier in this chapter. Impregnation, using
silane, is covered in a later Department of Transport document, BD43/90,
which supersedes the section on this topic in BD27/86.

The information contained in BD43/90 includes specifications for site
testing prior to impregnation as well as procedures and materials for silane
impregnation. Locations of test areas on bridges which have been in
service for more than 6 years are given, concentrating on those areas
likely to be subjected to direct salt spray or those areas below joints where
runoff containing salts from the deck above may have run down the
surface. Half-cell surveys and chloride content determinations are
specified. Guidance on which members are to be impregnated is given
in an associated Department of Transport document BA33/90.[8.13]

Material for impregnation is limited to monomeric alkyl
(isobutyl)trialkoxysilane with a minimum active content of 95 per cent
and zero hydrocarbon solvent content. The material is to be applied using
a circulated pump system with a nozzle pressure between 0.06 and
0.07 N mm^{-2}. The specification requires work to start at the lowest
level, working upwards using a continuous spray technique with an
application rate of 300ml m^{-2}. Two applications are to be made with
an intervening period of at least 6 hours. Certain restrictions are placed
on the ambient conditions during which impregnation can take place.
The shade temperature must be greater than 5 °C and the surface
temperature of the concrete under treatment must be less than 25 °C.

Cathodic Protection

The Concrete Society model specification for cathodic protection[8.6] is aimed at providing sufficient information to enable an engineer, who is not a specialist in cathodic protection but who is experienced in the preparation of contract documents, to draw up a detailed specification and bill of quantities for the design, installation, commissioning, and operation and maintenance of an impressed current cathodic protection system for a reinforced concrete structure. The specification given in the report is not intended to be reproduced word for word in a contract, but is intended to provide an overall framework containing alternative clauses which the specification writer can use to produce a contract-specific document by reference to guidance notes and the Society's companion report.[8.14]

The model specification contains seven sections as follows:

1. Introduction
2. Investigation and assessment
3. System requirements
4. Materials and equipment
5. Installation procedure
6. Commissioning
7. Operating and maintenance

The introductory section suggests that general details of the structure should be given, such as its location and condition, and the location of services available to the contractor, such as water and electricity.

The investigation and assessment section suggests tests which should be carried out on the structure. Some of these will obviously have been carried out in order to reach the decision to use cathodic protection prior to the production of contract documents, but there will often be a need to collect additional data during the implementation of the works. Clauses are given for tests particularly relevant to the efficacy of cathodic protection such as reinforcement continuity and concrete resistivity.

In the systems requirements section, details are given of the required current density in relation to reinforcement surface area, the layout of anode zones, the redundancy to be provided in the supply and anode systems to allow for possible cable breaks. Details of the control and monitoring systems are also given.

The materials and equipment section deals mainly with the properties of repair and overlay materials. Three types of material are listed for these applications: spray-applied materials, cast-in-place concrete, and poured or trowelled grouts and mortars. Requirements for workability, resistance to cracking and conductivity are given. A subsection on constituent materials deals with cement, admixtures and polymer content. It is noted that bonding agents or primers on the reinforcement must not be used. This section also gives details of testing. These include trial panels carried out before commencement of the main works, and coring of test panels during the course of the works of sprayed materials and cube testing for compressive strength of other materials. Testing of bond strength of overlays and repairs with the underlying concrete is also specified.

The detailed requirements for the various work processes in a cathodic protection contract are given in the installation procedures section. The section gives guidance on testing any previous repairs to make sure that the resistivities of the materials used are such that they should not impair the efficiency of the cathodic protection system. Requirements for removal of delaminated and honeycombed concrete, concrete repair and checking for and providing reinforcement continuity are also given. Alternative clauses set out the installation requirements for different anode systems — conductive coating, flame sprayed zinc and titanium mesh or conductive polymer wire. In the latter case, details of overlays of flowable concrete, render and gunite are also given.

There are two parts to the commissioning section. The first part deals with commissioning procedures such as circuit verification, adjustment of current output and performance verification. The second part deals with the documentation required at the time of commissioning. Details of the system, including as-built drawings, circuit diagrams and results of commissioning tests, are to be submitted in an installation and commissioning report. The contractor has also to compile an Operating and Maintenance Manual.

The operating and maintenance section of the specification describes three categories of inspection:

Functioning check
Performance monitoring
System review

The functioning checks are to be carried out on either a weekly or monthly basis depending on the complexity of the system and whether it is likely to be liable to vandalism or other forms of physical damage. Notes for guidance suggest 3-monthly intervals for performance monitoring and yearly intervals for system reviews.

The functioning check consists of determining that the power supply is switched on, and measurement of voltage and current supplied to each individual anode zone. Performance monitoring involves a visual inspection of the condition and integrity of the elements of the installed system and the protected structure along with measurement of steel/concrete potentials. Steel/concrete potentials are measured with cathodic protection current ON, under the instant-OFF condition and also after a period of depolarization. The system review consists of performance monitoring as described above, along with a functioning test of any embedded reference electrodes against a standard reference electrode and a review of all of the data recorded to date during performance monitoring checks.

An appendix to the specification lists items to be included in four sections of the bill of quantities. The list is generally drawn up in accordance with the *Civil Engineering Standard Method of Measurement*.[8.15] It is suggested that testing items should be included under the General Items section and measured by number. Items dealing with repair are listed under Concrete Repair. Removing existing repairs of high resistivity material and replacing with repair concrete is measured by item, as is installing replacement reinforcement including its continuity bonding. Surface preparation to receive the cathodic protection system is measured by area.

A special section is suggested for the installation of the various elements of the cathodic protection system, the installation of reference electrodes and monitoring probes, and commissioning and monitoring. The supply and installation of the cathodic protection system itself is listed under a single description measured by item. Provision of the power supply and commissioning with provision of report and operation manuals are also each listed separately and measured by item. The anode system installation is measured by area whereas monitoring probes, reference electrodes and the various functioning, performance monitoring and system review checks are measured by number.

A further special section contains the single item description for application of the overlay. This is measured by area.

References

8.1 Institution of Civil Engineers 1991 *ICE Conditions of Contract* 6 edn, The Institution of Civil Engineers, London

8.2 Joint Contracts Tribunal 1984 *Intermediate Form of Building Contract for Works of Simple Content* RIBA Publications, London

8.3 Federation Internationale des Ingenieurs — Conseils 1987 *Conditions of Contract for Works of Civil Engineering Construction* 4 edn, FIDIC, Lausanne

8.4 Concrete Society 1991 *Patch Repair of Reinforced Concrete Subject to Reinforcement Corrosion* Technical Report No 38, The Concrete Society, Wexham

8.5 Concrete Society 1979 *Specification for Sprayed Concrete* The Concrete Society, Wexham

8.6   Concrete Society 1991 *Model Specification for Cathodic Protection of Reinforced Concrete* Technical Report No 37, The Concrete Society, Wexham

8.7   British Standards Institution 1983 *Testing of Resin Composition for Use in Construction: Part 2 Method for Measurement of Compressive Strength* BS 6319, The British Standards Institution, London

8.8   Civieltechnisch Centrum Uitvoering Research en Regelgeving Research Committee B 35C 1990 *Determination of the Bond Strength of Mortars on Concrete* Recommendation 20, CUR, Gouda

8.9   Civieltechnisch Centrum Uitvoering Research en Regelgeving Research Committee B 35C 1990 *Concrete Repairs with Polymer-modified Cement Mortars* Recommendation 21, CUR, Gouda

8.10   Concrete Society 1981 *Guidance Notes on the Measurement of Sprayed Concrete* The Concrete Society, Wexham

8.11   Department of Transport Highways and Traffic 1986 *Materials for the Repair of Concrete Highway Structures* Departmental Standard BD27/86, Department of Transport, South Ruislip

8.12   Department of Transport Highways and Traffic 1990 *Criteria and Materials for the Impregnation of Highway Structures* Departmental Standard BD43/90, Department of Transport, South Ruislip

8.13   Department of Transport Highways and Traffic 1990 *Impregnation of Concrete Highway Structures* Departmental Advice Note BA33/90, Department of Transport, South Ruislip

8.14   Concrete Society 1989 *Cathodic Protection of Reinforced Concrete* Technical Report No 36, The Concrete Society, Wexham

8.15   Institution of Civil Engineers 1976 *Civil Engineering Standard Method of Measurement* Thomas Telford Ltd, London

Index

abrasion, 185
abseiling, 53
abutment, 43, 205
access, 53, 97, 145−6, 191−4, 196
acetic acid, 34
acid, 35, 179, 182, 185
  acetic, 34
  extraction, 131
  natural, 34
  organic, 34
acidic product, 183
acryllic, 156, 173, 180
adhesion, 174, 196
adhesive, 164
  tape, 83
admixture, 19−20, 25−6, 156, 199,
  208
  accelerating, 19, 24, 42
  air-entraining, 11, 19, 22, 24
  dispensing of, 23
  overdosing of, 23
  retarding, 19, 20
  superplasticizing, 20, 23
  underdosing of, 23
  waterproofing, 23
  water reducing/plasticizing, 19−24
aggregate, 13−14, 17−18, 22−3, 26,
  31−4, 36−8, 64, 70, 78, 81−2, 93,
  101, 104−8, 113, 126, 147−50,
  152−3, 160, 199, 201−2, 204
  dolomitic, 39
  grouted, 160
  lightweight, 155
  preplaced, 159
  sea-dredged, 18
  type, 105
aggregate/cement ratio, 150
air
  bubbles, 20
  entrainment, 22−3
  pocket, 205

void content, 105−6, 198, 200
alcohol, 171−2
alkali, 12, 18, 34−6, 38, 141, 185,
  188
  carbonate, 38
  content, 195
  hydroxide, 36
  metal, 35−6, 38
  reactivity, 35−8, 61, 107
  resistance, 202
alkaline
  condition, 166
  environment, 197, 199
  solution, 189
alkalinity, 6, 41, 187
alkali−aggregate reaction, 105, 189
alkali−carbonate reaction, 38−9
alkali−silica reaction, 36, 106, 138,
  141−3
alkali−silica reactivity, 107, 138−40,
  142−3
alkali−silicate reaction, 38
alkylisobutyltrialkoxysilane, 206
alkyl radical, 171−2
alumina, 6, 101
aluminate, 11
aluminium, 2, 12, 91, 177
  oxide, 3
aluminosilicate, 10−11
ambient temperature, 155
ammonia
  stearate of, 23
ammonium hydroxide, 105
anaerobic conditions, 34
anode, 43, 45, 167, 177−88
  cast iron, 179, 185
  conductive ceramic, 185
  conductive coating, 180, 182, 208
  conductive overlay, 185
  conductive polymer, 184, 208
  conductive resin, 185

distributed, 180
mesh, 180, 183
sacrificial, 177−8
slotted, 179−80
sprayed zinc, 184, 208
system, 181−2
titanium mesh, 182, 208
zone, 207, 208
anodic area, 43, 138, 176
antenna, 92
anticarbonation coating, 166
antifoaming agent, 199
appraisal, 139
arc spray, 184
area, cathodic, 43
Aristotle's lantern, 35
arris, 47
asbestos, 50
asphalt, 179
assessment, 124, 126, 129, 138, 145, 207
calculation, 125
attack, chemical, 46
attenuation, 93
Avonguard, 96

barium
chloride, 105
sulphate, 105
barrier coating, 156
basement, 50
batching, 148
bauxite, 3
Bayes' theorem, 118, 120−21
beam, 44, 47, 54, 57, 59, 64−5, 78, 89, 125−7, 141−3, 145, 165, 173, 205
precast prestressed, 45
bending, 140
moment, 57, 122
of resistance, 122
bentonite, 23
bill of quantities, 192, 196, 204, 207, 209
'bird's nest', 161
birefringent properties, 106
bitumen emulsion, 23
blastfurnace slag, 2, 11, 101
bleeding, 18, 23−4, 27−9, 205
blocks, 60
blowholes, 41, 169−70, 174, 195
blowpipe, 150
bond, 86, 140−1, 152, 155−7, 160, 182−4, 203
strength, 203, 208
bonding
agent, 208
aid, 152, 154−6, 158, 195
of external reinforcement, 164

brazing, 181
breaking out, 193−5
break-off test, 86
BRE internal fracture test, 83
brick, 60
panel, 125
brickwork, 204
bridge, 42−3, 53, 64, 141, 164, 175, 204
deck, 91, 179−80, 185, 187
brush, 169, 173
bubble
size, 106
spacing, 107
bughole, 169
building, 145
bulky repairs, 157
bund, 161
butadiene, 175
butyric acid, 34

calcite, 106
calcium, 2, 10, 12
aluminate, 11
carbonate, 2, 6, 38, 39
chloride, 19, 24−5, 42
formate, 19
hydrate, 12, 33
hydroxide, 6, 8, 9, 11−13, 19, 33, 35, 39, 106, 189, 201
nitrate, 19
oxide, 2, 46, 101−3
silicate, 11−12, 19, 33, 46, 173
hydrate, 13, 46
sulphate, 32−3
sulphoaluminate, 6
calibration, 76
capacity, 52−3, 143
capillary, 8, 9, 23, 31, 39, 46
action, 32
passage, 8, 9, 46
pore, 9, 104, 120, 170
porosity, 104
rise, 32
tube, 98−9
Capo test, 84, 85
carbon, 10, 11, 70, 85, 171, 184
dioxide, 2, 4, 34, 39, 40, 41, 101, 136, 156, 166, 170, 175−6, 201
fibre, 182, 186
carbonated
concrete, 95, 187
layer, 73, 95
sample, 101
carbonates, 10
carbonation, 36, 39−41, 43−4, 95, 105, 130−6, 145, 169−70, 175, 178, 194, 197−8
depth, 40, 54, 61, 95, 132−3, 176, 202

testing, 95
front, 39—41
car-park
    deck, 179, 185
    multi-storey, 180
car-parking structure, 43
cast-iron, 185
catalyst, 162
categorization system, 64
cathode, 45, 177
cathodic
    area, 43, 176
    polarization, 179
    protection, 166—7, 176—83, 185—7,
        193, 207—9
        current, 179, 186
        voltage, 186
    site, 43
cellulose fibre, 187, 188
cement, 6—9, 11—12, 19—21, 23, 28,
    35, 36, 40, 78, 101—2, 130—1, 133,
    136, 148—57, 199, 204
    air-entraining, 22
    changes, 7
    content, 8, 20, 22, 24, 61, 107,
        141, 199
        test, 102—4
    gel, 8
    grain, 101
    high alumina, 25, 45—6, 59, 68,
        70, 83, 102, 111, 126
    hydrate gel, 9
    hydrates, 156
    hydraulic, 168—9
    manufacture, 2
        dry process, 2, 4, 5
        Lepol process, 4
        semi-dry process, 2, 4—5
        semi-wet process, 2, 4—5
        wet process, 2—3, 5
    manufacturing processes, effects of
        changes, 7
    matrix, 130
    mortar, 145
    paste, 8, 9, 32—4, 46, 106, 107
    Portland, 2, 11—13, 17, 26, 45—6,
        83, 101—3, 129
        composition, 6
    properties, changes in, 7
    sulphate resisting, 17, 33—4, 101—3,
        136
    type, 105
        test for, 101—2
cementitious grout, 185
chalcedony, 36
chalk, 2, 4
characteristic
    bond strength, 198—9, 203
    compressive strength, 198—9, 201

dead load, 122, 127
flexural strength, 198—9, 201
live load, 122
load, 122
    strength, 113—22, 125
    value, 114, 124
chemical
    analysis, 104
    attack, 46, 105
    plant, 50
    test, 68, 93
        method, 101
    works, 168
cherry-picker, 53
chert, 36
chimney flues, 34
chloride, 16—18, 31, 33, 42—5,
    104—5, 111—12, 128—31, 133—6,
    145, 153, 166—7, 170, 172, 175,
    178, 187, 189, 206
    concentration, 105, 133
    contamination, 167
    content, 54, 61, 104, 128, 195
    extraction, 187—9
    ion, 175, 187—8
    penetration, 8
    profile, 129
chlorinated rubber, 173, 180
cladding unit, 54
Clam apparatus, 100
clamp, 34
classification system, 62, 65
classifying, 15
clay, 2—3, 10, 13, 16, 32, 35, 38
    minerals, 10, 38
clay, silt and dust content of
    aggregate, 17
cleaning, 146—7, 193
cling film, 70
clinker, 2, 4, 12—13, 102, 157
coal, 2, 10
coating, 164, 166—70, 173, 174, 176,
    182—3, 187, 189, 195, 198
    cement-based, 168
    thin, 169
cobalt-60, 95
coefficient of thermal expansion, 202
cohesion, 169
coke, 11, 179
cold joints, 26
coloidal silica, 23
column, 42, 44, 47, 50, 54, 56, 59,
    60, 65, 89, 125, 145, 150, 205
commissioning, 207—9
    report, 208
    test, 208
    testing work, 65
compaction, 56, 70, 80, 114, 147,
    149, 151, 157, 195—6, 198, 204

degree of, 203
composite
    beam and slab floor, 125
    construction, 126
composition, 197
compressed air, 163
compression, 98, 140, 151
    testing machine, 73, 101
    wave, 74
compressive strength, 54, 56, 68, 71,
    81, 141, 157, 201, 203−4, 208
    testing, 70, 202
concentrated load, 125
concrete
    flowing, 158
    low strength, 117
    pump, 148
    pumped, 159
condensed silica fume, 12
condition, 51, 52, 53
conditioning, 201
conditions of contract, 192
conductive
    asphalt, 185
    ceramic, 185
    coating, 180, 182, 208
    overlay, 185
    polymer, 180, 184
        anode, 184
        wire, 208
    resin, 185
conductivity, 93, 179, 208
confined space, 50
construction
    defect, 25
    joint, 151, 204
contact resistance, 91
contaminants, 174
contamination, 145, 147
contract document, 191, 194, 207
control
    specimen, 157
    system, 167, 207
copper, 89−91, 180−1, 184
    pipe, 161
    sulphate, 89−90
copper/copper sulphate half-cell,
    88−91, 137
coral, 35
core, 53, 57, 69−71, 73, 78, 86, 101,
    104, 107−8, 113−14, 117−18, 120,
    124−5, 138, 142, 151, 157, 164,
    174, 196, 204
    capping of, 70−1
    small diameter, 69
    testing, 69
coring, 58, 61, 78, 164, 208
corrosion, 16, 18, 19, 24, 37, 41−5,
    50, 61, 63, 129−33, 135, 137−8,

140, 145, 153, 160, 166, 169−70,
    172, 176−8, 181, 185, 186, 191,
    195, 197, 200
    inhibitor, 195
    pitting, 172
    potential, 89−90
    product, 45
    reaction, 41, 91
corrosive environment, 195
corrugated sheeting, 50
cover, 25−6, 45, 54, 63, 66, 86, 87,
    88−9, 92, 131, 132, 133−6, 150,
    154, 172, 178, 191, 192, 194, 197,
    199, 200, 204
covermeter, 57, 61−2, 66, 69, 86−9,
    95, 131, 194
crack, 26, 33, 36, 38, 40−1, 44−5,
    47, 61, 63−5, 70, 74, 78−9, 96,
    105, 107, 138, 142, 147, 160,
    161−4, 166, 172, 178, 187, 191
    depth, 79
    filling, 160
    injection, 146, 160
    mapping, 61
    pattern, 36, 37, 49, 57, 63, 78, 107
    plastic settlement, 28−30, 61
    plastic shrinkage, 26−8, 61, 161
    shrinkage, 151, 160
    system, 163
    width, 59, 107
cracking, 18, 38−9, 41, 44−5, 129,
    134−5, 141−2, 151−2, 160, 166,
    169, 178, 191, 208
    plastic, 24
cradle, 50, 53, 146, 194
crazing, 41, 64
creep, 47
crimping, 183
crossbeam, 172
crosshair, 96, 107
crosshead, 172−3, 205
cross linking, 172
crusher, 13−14
    cone, 14−15
crushing, 14−15, 63, 73, 204
    machine, 73
    strength, 69, 73
crystal growth, 31, 32
crystallization, 170
crystobalite, 36
cube, 68−70, 73, 78, 82, 151, 157,
    196, 200, 201−2
    crushing strength, 84
    strength, 84
    test, 197
cure, 163
curing, 9, 18, 26, 56, 78, 81, 85, 114,
    151, 155−7, 174, 195, 199, 204−5
    membrane, 151, 155

current, 178, 180–3, 186, 208
　density, 179, 181–2, 184, 206
　expansion, 138
cut section, 174
cyclone, 16
cylinder, 68, 73, 78, 82, 151

dairies, 34
damage, 145, 191
　class, 65
　classification, 65
data-logger, 62
data-logging facility, 173
daywork joint, 151
DC power source, 177, 182–4, 188
dead load, 123–4, 126, 153
debonding, 142, 183
defect, 61, 64, 147, 191–4
　construction, 25
　design, 25
　liability period, 194
deflection, 59, 60, 63–5, 127–8
deformation, 142
de-icing, 42
　salt, 43, 131, 179, 198, 203
delaminated area, 187
delamination, 45, 181
demec
　gauge, 96
　stud, 108, 142
density, 70, 74, 80, 124, 157, 204
depassivation, 130
depolarization, 187, 209
desalinated water, 35
desalination plants, 35
desert, 32, 42
design, 204
dessication, 46
detailing class, 140–1
deterioration, 51, 62, 65, 68, 76, 81,
　93, 101, 106, 141–2, 166–7, 191
　mechanism, 61
　processes, 112
dial gauge, 59–60, 96
diamond
　powder, 102
　saw, 102, 106
dicalcium silicate, 6
dielectric
　characteristic, 92
　constant, 93
diffusion, 129
　coefficient, 129
　resistance, 175
direct current, 178
dirt, 97, 147
discontinuity, 169
dissolved salts, 10
distortion, 64–5

domestic building, 126
drill dust, 93, 104
dry
　film thickness, 174
　mix process, 147–9
　pack, 155
drying out, 151
dummy gauge, 97, 98
durability, 8, 9, 22–4, 41, 62
dust, 97
　mask, 49
　sample, 93
　sampling, 61, 197

earth
　load, 123
　pressure, 123
earthquake, 49
eddy current, 86–7
effective
　depth, 125
　span, 127
efflorescence, 63
elastic modulus, 74, 76
electrical
　continuity, 181
　potential, 41, 44–5, 89
　resistance, 186
　　gauge, 96
　　probe, 186
　resistivity, 178
electrocatalytic role, 183
electrochemical, 43, 45, 89
　process, 176
　protection, 156–7
electrode, 89
　potential, 173, 181
electrolyte, 89, 187
electrolytic
　process, 182, 189
　technique, 166
electromagnet, 97
electromagnetic
　device, 86
　pulse, 93
electro-osmosis, 189
E-log I determination, 187
environment, 166
environmental condition, 133, 135,
　185, 198
epoxy, 156–7, 175
　adhesive, 96, 164
　mortar, 182
　resin, 102
Epsom salt, 33
equivalent sodium oxide content, 35
erosion, 49
estimated *in situ* cube strength, 70
etching, 101–2

ethyl alcohol, 95
ettringite, 6−7, 33, 106, 136
evaporation, 26, 27, 42
excess voidage, 70
excessive deflection, 57
existing records, 52−3
exothermic reaction, 163
expanded
    mesh, 180
    metal, 182
expansion, 108, 142
    index, 107, 138−9, 141
        current, 107
        potential additional, 108
explosion, 49
exposed aggregate finish, 155
exposure condition, 74, 111, 133
extraction of aggregates, 17

façade, 41, 145−6
failure, 64
fairing coat, 169, 197
farm silage, 34
feather edge, 152, 195
ferric hydroxide, 34, 44
ferrite, 102
ferrosilicon alloy, 121
ferrous
    ions, 43
    oxide, 101
    sulphate, 34
    sulphide, 34
fibre, 169, 188−9
Fick's law, 129
field
    sheet, 46, 62
    telephone, 59
Figg permeability apparatus, 99
filler, 152, 169, 195, 199
film, 169−70, 173
    free, 176
    thickness, 169, 174
        dry, 174, 185
        wet, 174
film forming material, 169
filter funnel, 98
final set, 155
finishing, 155
fire, 49−50, 59, 64, 74, 145, 165
    authority, 53
    damage, 46, 61, 64−6, 105
fire-escapes, 50
fixed price, 191
flakiness index, 17
flash coat, 151
flat, 126
flaws, 74
flexural
    capacity, 142

strength, 156, 201
    test, 202
flexure, 125
floor, 65, 126
    panel, 65
    slabs, 45, 54, 59
flow test, 205−6
flowing concrete, 205
flues, chimney, 34
fly ash, 2
foaming, 199
foil strain gauge, 97
formaldehyde, 20
formed surface, 74
formwork, 28−30, 42, 45, 74
foundations, 61
free expansion, 138
freeze/thaw, 22−3, 31
freezing, 151
frequency, 93
    distribution, 117, 120
frost, 8, 24, 61
    attack, 74
    damage, 140
    resistance to, 198, 203
functioning check, 208−9

galvanic corrosion, 181
galvanizing, 157
gamma ray, 95−6
gas, 167, 169−70
    chromatography, 176
gauge
    dummy, 97−8
    length, 97, 108
    strain
        acoustic, 97
        demec, 96
        electrical resistance, 96
        mechanical, 96
        vibrating wire, 96−7
gel, 8, 36, 38, 106
geology, 105
ggbs, 12
gibbsite, 46
glass fibre, 182
Glaubner's salt, 33
grading, 13−14, 17, 18
graphite, 180, 182, 185
gravel, 12, 14, 16−17
grease, 75, 97
grid, 66
grinding of cores, 70
grit blasting, 153−4, 164, 174, 194−5
ground granulated blast-furnace slag,
    10, 12, 36, 103, 105−6
groundwater, 32−4, 45
grout, 160
    cementitious, 185

grouting of preplaced aggregate, 158, 160
gunite, 147, 208
gypsum, 2, 5–6, 32, 33
    plaster, 71
gyratory crusher, 14–15

HAC, 45, 46
Hach test kit, 104
half-cell, 61–2, 66, 89–91, 137
    copper/copper sulphate, 186
    potential, 89, 136, 137
        mapping, 172
    silver/silver chloride/potassium
        chloride, 186
    survey, 206
halogen content, 198, 202
hammer, 71
harbour, 42
hardener, 162
hardness, 71, 73–4
health and safety, 198
heat evolution, 8
high humidity, 151
highway, 23, 31, 42
    bridge, 131, 137
    structure, 205
histogram, 114–16, 131, 132
honeycombing, 26, 63, 70, 95
house, 126
humid conditions, 154
humidity, 181
hydrated
    cement matrix, 101
    concrete, 172
hydration, 8, 23, 46, 174
    heat of, 11, 13
    products, 101
    reaction, 6, 8, 12–13, 104
    temperature, 12, 31
hydraulic
    cell, 86
    jack, 84
hydrocarbon solvent content, 206
hydrochloric acid, 91, 102, 105
hydroflouric acid, 102
hydrophobic
    agent, 172
    compound, 170
    material, 170
    surface, 171
hydroxycarboxylic acid, 19–20
hydroxyl ion, 19, 38, 43, 187, 189
hypodermic needle, 51, 98–9

I beam, 141
impact, 49
impactor, 14–15
imposed load, 123

impregnation, 166, 204, 206
impressed current, 177–8
indirect
    method, 113
    test, 113
industrial process, 112, 146
inert filler, 163
initial surface absorption test, 98
injection, 160, 162–4
    port, 161–2, 164
insoluble residue, 101
initial survey, 65
in situ tests, 62
inspection, 61, 111, 194, 196
    visual, 196
installation, 207, 209
instant-OFF potential, 209
instrumentation, 60, 142
insulation, 50
interpretation of results, 111
invar, 60, 97
investigation, 111, 191, 207
iridium-192, 95
iron, 2, 10–11, 43
    ore, 11
    oxide, 2–3, 6

jaw crusher, 14–15
joist, 126–7
joist and hollow block construction, 126

kinetic energy, 81

laboratory test, 68, 142
    methods, 101
lactic acid, 34
ladder, 53
laitance, 85, 97, 151
lamination, 45
lap, 153
    length, 154
lapping, 106
large
    deflection, 49
    pour, 68
latex, 156
leaching, 106
lead foil, 96
leakage, 126
leptospiral jaundice, 51
letter box shutter, 158–9
levelling
    coat, 169
    mortar, 152
life requirement, 145
lightweight aggregate, 13
lignosulphonate, 19–20, 22
lime, 173

lime-saturated water, 71
limestone, 2—4, 11, 23, 34, 38
  Ballidon, 103
  dolomitic, 38—9
  siliceous, 36
limit state, 122
Limpet test, 85
linear traverse method, 106—7
link, 89, 142
liquid, 169
  soap, 75
lithium oxalate, 19
lithophago pirhuana, 35
load, 123, 179
  combination, 122—4
  factor, 122
  live, 153
  restriction, 142
  test, 57—9, 111, 127, 140
  testing, 141—2
  transfer, 124
  variation factor, 122, 124
load carrying capacity, 49, 51, 56, 68, 141
loading, 60
logarithm, 114, 116
  natural, 116—17, 120
log-normal distribution, 114, 121
Lok test, 84
long chain compounds, 171
loss on ignition, 101
lump sample, 61, 101, 104
lump sum form of contract, 192

macrocracks, 37
magnesium, 10, 12, 177
  carbonate, 38
  hydroxide, 33
  oxide, 38, 101
  perchlorate, 104
  sulphate, 33
magnetic
  field, 86
  induction, 86
maintenance, 145—6
  inspection, 141—2
  liability, 167
  period, 194
  strategy, 139
maisonette, 126
manometer, 99, 100
manufacture, 197
map cracking, 61
marine
  aggregate, 36
  borers, 35
  environment, 154
  gravel, 14
  structures, 32, 42

marl, 3
martesia striata, 35
masonry, 125, 167—8
materials, 193, 195
matrix, 34, 101, 104, 156, 173
measuring
  coil, 87
  head, 88
mechanical properties of aggregates, 17
mechanical
  coupler, 154
  gauge, 96
mean, 114
  strength, 55, 113, 119—20
  value, 54, 119, 120
measurement, 204
melamine formaldehyde, 20
mesh, 182—3, 188
metal weight, 60
methacrylate, 175
  methyl, 175
  polyisobutyl, 175
method of measurement, 196
microcracking, 36, 156
microcracks, 38
micrometer, 101
microscope, 102, 164
  metallographic, 106
  metallurgical, 106
  method, 101, 105
  polarizing petrological, 106
  stereoscopic, 105—6
microscopic examination, 105
microscopical test method, 101
microsilica, 2, 10, 12—13
milk, sour, 34
mineral phases, 102
mix, 150
  proportions, 105—6
mixing, 195
  head, 163
modelling clay, 98, 161
modulus of elasticity, 74, 141
moisture, 32, 34, 37, 41—4, 68, 74, 89, 97, 132, 138, 156, 161, 170, 172—3
  condition, 72—3, 82
  content, 60, 72, 78, 80, 93, 149, 199
  scavenger, 157
monitoring
  equipment, 180
  probe, 209
mortar, 126, 152, 156, 157, 172, 198, 199, 200—1, 203, 204—5
  block, 175, 176
  cement, 145, 155
  cube, 157

epoxy, 182
levelling, 152
polymer, 152, 154—5, 179
    modified, 152
        cementitious, 155, 157, 183, 197
repair, 156, 205
molluscs, 35
moment, 122, 126
monitoring, 111
performance, 145
monocalcium aluminate decahydrate,
    46
mould, 158

naphthalene, 23
national standards, 65
niobium, 179, 185
nitric acid, 104
nitrogen, 104
non-destructive
    method, 125
    test, 68, 69
normal
    curve, 115
    distribution, 54, 113—14, 116, 119

office building, 180
olive oil, 22
opal, 36
operating and maintenance manual,
    208
operator bias, 73
overalls, 50
overlay, 86, 167, 179, 180, 183—4,
    207—9
    cementitious, 183—5
oxygen, 26, 34, 37, 41, 43—5, 138,
    166, 170—1, 179, 187
    diffusion coefficient, 176
oxygen/acetylene gun, 184
oxygen/propane gun, 184

pachometer, 86
paint, 187
    film, 175
partial factor, 123—4, 127
partial load factor, 125
partial safety factor, 123—6
particle shape, 13, 17—18
partition, 124
passivating effect, 157
passivation, 43
passivity, 6
patch repair, 145—6, 152—3, 155,
    158, 181, 193, 196
    system, 152
penetration
    depth, 172
    resistance, 80—2, 204

test, 58
percussion drill, 93—4
performance monitoring, 208, 209
permeability, 8, 9, 23—4, 98, 156,
    169, 173, 175—6, 179, 187
petrographic
    analysis, 136
    examination, 61, 101, 105
    technique, 105—6
petrography, 105, 107
petroleum jelly, 23
PFA, 10—12, 101
pH, 33, 39—42, 95, 189
phenolphthalein, 95, 176
phyllite, 38
piano wire, 150
piddock, 35
pier, 42, 205
pigment, 169
pinhole, 169, 174
pipe, 35
pipeline, 177
pit, 34, 45, 153
placing, 205
planning an investigation, 49
plaster, 147
    of Paris, 76
plastering, 169
plastic
    bag, 70
    barrel, 60
    behaviour, 13
    cup, 94
    settlement, 31
    shrinkage, 31
    state, 155—6, 169
plasticizer, 150, 169
plasticizing effect, 156
platform, 146
    self-elevating, 146
platinum, 179
point
    counting, 106
    load, 142
poisonous gas, 50
Poisson's ratio, 74
polarization, 186
polarized light, 106
polished
    sample, 101
    section, 106, 164
    specimen, 106—7
polishing, 106
polyester
    putty, 161
    resin, 102
polymer, 154—7, 169, 184, 199, 200
    cement slurry, 154
    content, 199, 208

dispersion, 173
emulsion, 154
latex, 156
mortar, 145
putty, 162
redispersible powder, 156
repair mortar, 154
polymerization, 170, 172
polysiloxane, 172
polythene
sample bag, 93
polyurethane, 173
ponding, 170
poor compaction, 95
pop-outs, 18, 32, 37, 61
pore, 8, 9, 23, 31–3, 38–9, 41, 46,
91, 167–8, 170, 173
blocker, 23
filler, 152, 169
fluid, 179
liner, 23, 167–71
solution, 38–44, 131, 167–8,
170–1, 173
stopper, 169
structure, 199
system, 167
water, 33, 36
port, 161, 163
portlandite, 6
positive skew, 114–15
potassium, 6, 10, 35
hydroxide, 36, 102
oxide, 35
silicate, 36
potential, 173, 177, 187
contour, 138
decay, 186–7
difference, 177
shift, 187
potential additional expansion, 138
Pourbaix diagram, 41–3
powder, 156
latex, 156
power supply, 180–1, 187, 208
pozzolana, 106
pozzolanic
action, 13
material, 2, 11
properties, 10
reaction, 11–12
precasting, 42
preliminary inspection, 63
pressure, 161
positive, 163
prestress, 141–2
prestressed concrete, 141
prestressing steel, 47
primary member, 124
primer, 195, 208

priming agent, 203
prism, 200–4
probability, 119–21, 124, 137
distribution, 117
paper, 117–18
scale, 129, 130
probe, 186
built-in, 173
electrical resistance, 186
macro-cell, 186
pick-up, 186
processing of aggregates, 17
progressive collapse, 124
properties of aggregates, 17
proportioning, 148
propping, 57, 111, 195
protection, 111, 166–8, 196
criterion, 179
system, 186
protective
coating, 147, 152
treatment, 174, 197
pull-off
strength, 157, 196
test, 58, 82, 85
testing, 157
pull-out test, 58, 82–3
pulse
generator, 74, 75
velocity, 76–80, 113
equipment, 76
pulverized fuel ash, 10, 13, 36, 103,
105, 106
pumping, 158
purlin, 126
pyrites, 34
iron, 34

quality control, 114, 126
Quantab titrator strip, 104
quartz, 12, 36
sand, 157

radar, 92–3
radiography, 57, 95
random number, 55
rate of sampling, 53
reaction
product, 106, 189
rim, 105
reactive aggregate, 37
rebound, 149–51, 204
hammer, 71–4, 82, 124
number, 72–4
test, 73
reconstruct, 112
record office, 53
recording results, 62
recovery, 127

redundancy, 207
reference electrode, 186, 209
    copper/copper sulphate, 186
refurbishment, 64
regression, 118
rehabilitation, 191–3, 196
    contract, 191
reinforcement, 16, 18–19, 26, 28, 33,
    37, 41–5, 47, 49, 54, 56–9, 61,
    63–6, 68–70, 80, 83, 86, 88–90,
    92, 95, 97, 112, 123, 125, 129–35,
    138–41, 145, 150, 153–8, 160–1,
    166, 169, 172, 178–81, 186–9,
    191–2, 194–5, 197–200, 204,
    207–9
    detailing class, 140
    primer, 152, 154
    protection, 152, 156, 157
    supplementary, 153
reinforcing
    bar, 86, 87
    drawing, 88
    steel, 137
reinstatement, 154, 193, 195, 200
relative humidity, 26, 40, 68, 107,
    108, 136, 173, 175, 201
remeasurement, 191–2
remedial work, 173, 191, 194
render, 167–8, 208
    cementitious, 180
    polymer bound, 168
rendering, 167–9
    cementitious, 174
renovation, 166
    technique, 145
repair, 63, 64, 86, 111–12, 145–7,
    166, 191, 196–7, 200, 204, 205,
    208–9
    contract, 146, 192
    method, 166
    system, 153
    technique, 65, 145
    type, 65
replenishment of alkalis, 187–9
resin, 106, 160–4, 184
    injection, 160, 162, 164
resistance, 97, 186
resistivity, 91–2, 173, 181–2, 208
resolution, 93
retaining wall, 29
rock, 13–14, 32, 35
roller, 169, 173
roof, 31, 46, 126
runway, 31
rust, 143
    staining, 34

safety, 49, 112
    boots, 50

clothing, 50
equipment, 50
factor, 140
hats, 50
procedures, 50
salt, 18, 31–2, 42, 43, 93, 167, 175
    crystals, 31–2
    spray, 206
    weathering, 31–2, 37, 61
sample, 54
    bag, 94
    size, 55
sampling, 54, 62
    error, 54–5
sand, 13, 16, 26, 36, 60, 148–50,
    155–6
    beach, 18
    blasting, 147
    pocket, 150–1, 204
    siliceous, 70
satellite tower, 53
SBR, 156
scaffold, 60
    light-weight, 53
scaffolding, 50, 59, 146, 194
scatter plot, 118, 133, 134
schedule of rates, 196
Schmidt hammer, 71–2
scissor lift, 53
screed, 150
screen, 124
screening, 14–15
scrubbing, 147
sea, 42
    urchins, 35
    water, 18, 33, 36, 178
sealer, 167, 173
search head, 86–8
secondary member, 124
segregation, 18, 24, 205
seismology, 92
self-elevating platform, 194
sensitivity factor, 122, 124
setting time, 164
settlement, 49
    plastic, 28, 31
sewage, 34
    treatment plant, 168
sewer, 50
shear, 122, 125, 140
    capacity, 142
    movement, 142
    reinforcement, 141–3
    strength, 141
shell content of aggregate, 17
shotcrete, 147
shrinkage, 18, 30–1, 156, 160, 163,
    198, 202
shutter, 158

shuttered repair, 158
sieve, 101
silane, 171−2, 174−5, 206
silanol, 171, 172
silica, 6, 12−13, 35−6, 38, 46
  colloidal, 23
  fume, condensed, 12
  soluble, 101, 103
silicate, 11, 38, 102, 172−3
  minerals, 38
silicoflourides, 173
silicon, 2, 12, 171, 173
  carbide, 102
  oxide, 2
silicone, 171
  resin, 171−2
  rubber, 98
silo, 34
siloxane, 171, 174
silver, 89
  chloride, 89
silver/silver chloride half-cell, 89
site
  cathodic, 43
  environment, 140
  mixing, 205
  testing, 206
skew distribution, 121
slab, 26, 29, 77, 88−9, 94, 125, 145,
  158, 161, 165
  thin, 124
slag, 12−13
slate, 38
slump, 149
sodium, 6, 10, 35
  hydroxide, 33, 36, 91
  oxide, 35
  silicate, 36
  stearate, 23
  sulphate, 33
soffit, 74, 94, 145
soil, 32−3, 45, 178, 179, 180
solvent, 172
spall, 44
spalling, 46−7, 49, 61, 63−5, 81,
  129, 166, 178, 191
Spearman's rank, 118
specification, 173, 192−4, 204−5,
  207, 209
splice, 154
spray, 169, 173
sprayed concrete, 145−7, 149, 150,
  157, 193, 204−5
  code of practice, 147
  method of measurement, 147
  specification, 147
spraying, 183, 204
staining, 49
stainless steel, 150

plate, 184
standard deviation, 54−6, 113−15,
  119−20
  population, 54
  sample, 55, 201, 203
statistical theory, 54
steel, 42, 47, 69, 75, 125, 187
  float, 151, 155
  mesh, 188
  plate, 164
  structure, 177
  trough, 205
  wool, 91
steel/concrete potential, 209
stiffness, 73, 164
stirrup, 89
stone, 13, 26
storage, 195, 205
strain, 96, 97, 108, 129
  gauge, 60, 85
    acoustic, 97
    demec, 96
    dummy, 98
    electrical resistance, 96
    mechanical, 96
    vibrating wire, 96−7
strand, 47
strength, 7, 8, 13, 17, 20, 22, 23−4,
  27, 46−7, 54, 56−9, 68, 71−4, 76,
  78, 80−2, 84−6, 113−20, 124−6,
  140, 141, 149, 151, 164, 173, 195,
  204
  correction factor, 71
  development, 83
  result, 119
  test, 61, 69, 113, 124
  testing, 69, 204
  very high, 13
strengthening work, 142
stress, 164
  redistribution, 122
  wave, 73
stress−strain
  curve, 125
  relationship, 73
structural
  adequacy, 58
  appraisal, 140
  assessment, 108, 121, 124
  capacity, 56, 57
  failure, 141
  integrity, 63
  performance factor, 122
  severity rating, 140−2
structure, car-parking, 43
student's t, 55
styrene butadiene, 156
substrate, 150
suction, 154

sugar, 19
sulphate, 6, 12, 31−3, 45, 136
    attack, 8, 32−3, 61, 68, 74, 106
    content, 61, 105, 136
sulphide, 10
    content, 103
sulphoaluminate, 7
sulphur, 70
    mortar, 71
sulphuric acid, 34
surface
    cleaning, 193, 196
    coating, 167
    defect, 169
    finish, 204
    preparation, 195
    protection, 193, 195
    treatment, 166−7, 173, 174
surface
    active agent, 20
    cleaning, 197
    hardness test, 58
    layer, 78−80
    preparation, 209
    staining, 63
survey, 51−2, 69, 111, 147, 191,
    193−4, 197
    work, 194
syringe, 99
system review, 208−9

tallow, 22
T-beam, 125
tell-tale, 96
temperature, 26, 68, 97−8, 136, 163,
    173, 174, 181, 195, 200−1, 206
    ambient, 200
    high, 68
    low, 68, 199
tendon, 142
ten percent fines, 17
tensile strength, 140−1, 156
tension, 97, 98, 160, 164
test, 65, 68−9, 157, 196, 207
    execution, 200
    load, 127
    panel, 151, 204, 208
    primary, 200, 202−3
    result, 196
test specimen
    composite, 200−1, 203
testing, 61−2, 65, 69, 157, 164, 186,
    193, 195−8, 200, 204, 208−9
    machine, 69
    technique, 68
tetracalcium aluminoferrite, 6, 7
Thames valley
    flint, 103
    sand, 142

thermal
    conductivity, 46
    cracking, 8
    expansion coefficient, 202
    extension, 47
    movement, 60
        early, 28, 61
    shock, 47
thin section, 106
thixotropic gel, 163
time-related
    element, 197
    item, 196
titanium, 180, 182, 185
    mesh, 182−3, 188
    plate, 184
titration, 104
titrator strip, 104−5
tolerance, 204
torque meter, 84
total
    alkali content, 35
    expansion, 138, 140
tower, 146
traffic spray, 174
transducer, 74−9
    configuration, 76
        direct, 76−7
        indirect, 76−8
        semi-direct, 76−7
    exponential probe, 76
transformer-rectifier, 180−1
transmission length, 142
trial, 193
    panel, 208
    repair, 194, 196
    sample repair, 196−7
tributyl phosphate, 21
tricalcium aluminate, 6−7, 33−4, 102,
    136
    hexahydrate, 46
tricalcium silicate, 6, 13
tricalcium sulphoaluminate, 33
trichloroethane, 104
tridymite, 36
trowel, 155
trowelled surface, 74
tunnel, 149
twinning, 102
two-way radio, 59
tying wire, 181, 187

ultimate limit state, 123, 127
ultrasonic pulse, 77, 80
    velocity, 74, 76, 78, 82, 113, 124,
        141
    test, 58
ultrasound, 74
    pulse, 74

ultraviolet radiation, 170
unhydrated cement, 95
uniformly distributed load, 125
urethane, 175
  moisture cured, 175

vacuum, 104, 163
  dessicator, 104
  drill, 94
  pump, 99
vandalism, 208
vapour transmission, 175
velocity of sound, 74
velocity−strength relationship, 76
vermiculite, 38
vibrating wire gauge, 96−7
void, 28, 32, 70, 72, 74, 80, 92, 105,
  107, 163, 204
voltage, 180−1, 188, 208
voltmeter, 90

walk-over survey, 51−3
wall, 50, 54, 65, 139
warehouse, 50
washing of aggregates, 16
water, 6, 16, 18, 28, 30, 39, 60, 69,
  91, 98−100, 104, 108, 148−51,
  154−6, 170, 179, 199, 207
  absorption, 175
  combined, 104
  content, 104
  extraction, 131
  gauging, 150
  jet
    high pressure, 153
  jetting, 147, 194
  load, 123
  penetration, 198, 202
  pressure, 123
  repellant, 171
    properties, 172
    surface, 172

table, 42
  vapour, 176
water/cement ratio, 7−9, 22, 40,
  78, 107, 126, 135−6, 148, 150
waterproof clothing, 50
wearing course, 179
weather
  adverse, 200
  rainy, 200
weathering, 32
  agent, 170
wedge anchor bolt, 83
weight, 108
Weil's disease, 51
welding, 153, 183
Wenner array, 91
wet
  environment, 141
  mix process, 147−9, 151
  spraying, 197
wetting
  agent, 91
  characteristic, 170
Wheatstone bridge, 97
wind, 26
  load, 123
wire, 47
  brushing, 153
  strain gauge, 97
wooden float, 151, 155
workability, 8, 18, 20, 22−3, 26, 30,
  207
workmanship, 56

yield, 140
  line, 125
  point, 125

zinc, 91, 184−5
  sprayed anode, 184